T0138618

Modeling and Design of Electromagnetic Compatibility for High-Speed Printed Circuit Boards and Packaging

Modeling and Design of Electromagnetic Compatibility for High-Speed Printed Circuit Boards and Packaging

Xing-Chang Wei

CRC Press
Taylor & Francis Group
Boca Raton London New York

CRC Press is an imprint of the
Taylor & Francis Group, an **informa** business

CRC Press
Taylor & Francis Group
6000 Broken Sound Parkway NW, Suite 300
Boca Raton, FL 33487-2742

Library of Congress Cataloging-in-Publication Data

Names: Wei, Xing-Chang, author.
Title: Modeling and design of electromagnetic compatibility for high-speed printed circuit boards and packaging / Xing-Chang Wei.
Description: Boca Raton : CRC Press, Taylor & Francis Group, [2017] | Includes bibliographical references and index.
Identifiers: LCCN 2016053999 | ISBN 9781138033566 (hardback : alk. paper) | ISBN 9781315305875 (ebook)
Subjects: LCSH: Printed circuits--Design and construction. | Electronic packaging. | Electromagnetic compatibility.
Classification: LCC TK7868.P7 W44 2017 | DDC 621.3815/31--dc23
LC record available at https://lccn.loc.gov/2016053999

Visit the Taylor & Francis Web site at
http://www.taylorandfrancis.com

and the CRC Press Web site at
http://www.crcpress.com

Contents

Preface

A high-speed circuit is the base of contemporary information and communication technology (ICT) and consumer electronics. Our modern life is heavily dependent on the functioning of high-speed circuits developed for various purposes. Therefore, the electromagnetic compatibility (EMC) among various circuits becomes very important. Because of ever-increasing clock frequencies and their harmonics, the electromagnetic field spectrum on the printed circuit board (PCB) and inside the electronic package extends to several tens of gigahertz. With decreased wavelengths, the electromagnetic wave phenomenon on the PCB and inside the package is obvious. This makes the signal spread over the entire PCB and package instead of confirming along the traces. At the same time, the three-dimensional integration including the through silicon via (TSV) based system-in-package (SiP) greatly increases the circuit density and complexity. Such a high frequency and a high density become a big challenge for the EMC control of high-speed circuits.

This book presents EMC modeling and control methods based on the author's many years' of research. The emphasis of this book is placed on two essential passive components of a high-speed circuit: the power distribution network and the signal distribution network. The field-circuit hybrid modeling and simulation methods of these passive components are discussed in detail. In addition, this book also explores the applications of novel structures and materials, including high-impedance surfaces and graphene films, in the EMC design, where the traditional bulk EMC materials cannot be used.

The signal integrity (SI), power integrity (PI), and electromagnetic interference (EMI) are three major EMC issues related to a high-speed circuit. For each of them, there are lots of EMC problems. Because of this diversity, the EMC is always analyzed and designed on a case-by-case basis. An EMC beginner usually finds it difficult to know where to start. However, behind all EMC problems is the behavior of the electromagnetic field in different environments. The study of the electromagnetic field characteristics is the key to better understand EMC problems and their control methods. The electromagnetic modeling is very important and is the best way to accurately get an insight into the electromagnetic phenomena. From that insight, some general theories can be obtained from disordered EMC phenomena. With the advancement of the computing technology, EMC design has changed from the *cut and try* to the engineering art in recent years. Most of the potential EMC problems can be predicted with a remarkable accuracy through electromagnetic simulation. This book therefore provides a detailed description of the electromagnetic modeling of the EMC problems.

The author hopes that this book will serve a basis for further progress in the high-speed circuit EMC for both academic research and industrial applications.

Acknowledgments

Some materials presented in this book are from the research works of my team. I acknowledge many of those individuals, including Decao Yang, Jun Li, Da Yi, Yufei Shu, Lingsong Zhang, Xiaojuan Wang, Dong Wang, Hanqin Ye, Jianbo Zhang, Xin Wei, and Liang Gao, all from Zhejiang University, Hangzhou, China. I also thank Lili Yang and Weiying Ding for proofreading the draft. It is a great pleasure to work with my PhD and master students. The research work in this book is also supported by National Science Foundation of China (61274110).

About the Author

Xing-Chang Wei is a senior IEEE member. He received a PhD degree in electrical engineering from the Xi'an University of Electronic Science and Technology, Xi'an China, in 2001. From 2001 to 2010, he worked at the A*STAR Institute of High Performance Computing, Singapore, as a research fellow, then as a senior research engineer, and finally as a research scientist. He received the 2007 Singapore Institution of Engineers' Prestigious Engineering Achievement Award for his contribution toward the development of a novel electromagnetic compatibility (EMC) measurement facility. In 2010, he joined Zhejiang University, Hangzhou, China, as a full professor, and received the New Century Excellent Talents Award from the Chinese Ministry of Education.

His main research interests include power integrity and signal integrity simulation and design for high-speed printed circuit boards, through silicon via analysis, and the development of fast algorithms for computational electromagnetics. He has more than 10 years' research experience of the EMC modeling and design of the high-speed printed circuit boards and packaging. In this field, he has authored or coauthored more than 50 papers published in *IEEE Transactions* and IEEE international conferences. He was the cochair of the Technical Program Committee (TPC) of the 2010 IEEE Electrical Design of Advanced Packaging and Systems Symposium, and severed as a TPC member and the program chair of the Asia-Pacific Symposium on Electromagnetic Compatibility from 2010 to 2013. He also delivered many workshops on signal integrity, power integrity, and EMC of high-speed circuits at the Asia-Pacific Symposium on Electromagnetic Compatibility and various universities in 2011, 2012, and 2015.

He conducted several EMC-related projects founded by Singapore and Chinese governments. He also had industrial projects with Huawei Company in China to solve its practical EMC problems in its high-speed IT products.

Acronyms

AP	application processor
APC-7	amphenol precision connector
CE	conducted emission
CM	common mode
CMF	common mode filter
CMOS	complementary metal–oxide–semiconductor
CPW	coplanar waveguide
CS	conducted susceptibility
CVD	chemical vapor deposition
DE	differential evolution
Decaps	decoupling capacitors
DGS	defected ground structure
DI	deionized
DM	differential mode
EBG	electromagnetic bandgap
EFIE	electric field integral equation
EMC	electromagnetic compatibility
EMI	electromagnetic interference
EMS	electromagnetic susceptibility
ESL	equivalent series inductance
ESR	equivalent series resistance
FDCL	frequency-dependent cylinder layer
FDTD	finite-difference time-domain
FEM	finite element method
FET	field-effect transistor
FEXT	far end cross talk
FSS	frequency selective surface
FTO	fluorine-doped tin oxide
GA	genetic algorithm
GS	ground-signal
MFIE	magnetic field integral equation
MoM	method of moments

MOS	metal–oxide–semiconductor
NEXT	near end cross talk
HIS	high-impedance surface
IC	integrated circuit
ICT	information and communication technology
IEEC	integral equation equivalent circuit
IL	insertion loss
IMD	intermetal dielectric
ITO	indium tin oxide
ITRS	international technology roadmap for semiconductors
PCB	printed circuit board
PDMS	polydimethylsiloxane
PDN	power distribution network
PEC	perfect electric conductor
PEEC	partial element equivalent circuit
PET	polyethylene terephthalate
PGGs	power–ground grids
PGPs	power–ground planes
PI	power integrity
PMC	perfect magnetic conductor
PMMA	polymethyl methacrylate
RDL	redistribution layer
RE	radiated emission
RS	radiated susceptibility
SDN	signal distribution network
SE	shielding effectiveness
SEM	scanning electron microscope
SerDes	serializer-deserializer
SGS	signal–ground–signal
SI	signal integrity
SMA	subminiature A
SSN	simultaneous switching noise
TEM	transverse electromagnetic
TGV	through glass via
TL	transmission line
TM	transverse magnetic
TSV	through silicon via
UWB	ultrawideband
VNA	vector network analyzer
VRM	voltage regulator module
XT	cross talk

Chapter 1

Electromagnetic Compatibility for High-Speed Circuits

Information and communication technology (ICT) and consumer electronics have been driving the semiconductor industry to integrate more and more circuits into one single package or printed circuit board (PCB). At the same time, the voltage supply level is continuously reduced with the ever-increasing working frequency. Typical operation frequencies for DDR4 range from 1.5 to 4 GHz at 1.1 V supply voltage [1]. The high-density, high-speed, and low-noise margin makes the electromagnetic compatibility (EMC), including the signal integrity (SI), power integrity (PI), and electromagnetic interference (EMI), critical for the successful design of a high-speed integrated system. According to 2013 International Technology Roadmap for Semiconductors (ITRS) report [2], "increasing noise effects, such as cross talk and power–ground bounce, decrease noise and timing margins and increase circuit susceptibility to defects." Therefore, there is a great demand for the efficient EMC modeling and design for high-speed PCBs and advanced packaging.

Considering the increased clock frequency and its harmonics, the interesting electromagnetic spectrum on the PCB and inside the package will cover several tens of GHz. With their ever-decreasing wavelength, the electromagnetic wave phenomenon inside small circuit structures cannot be ignored. With increasing frequencies, the EMC research effort moves from the PCB level to the package level or even the chip level. In the past, when the working frequency was below gigahertz, the EMC problems could be solved by using the low-frequency circuit theory due to the electrically small dimension of the structure under study. For many years, people

1

had developed lots of low-frequency EMC modeling and design methods. However, now we face a new EMC challenge for current gigahertz and future terahertz high-speed circuit designs, where we need to use the electromagnetic field and microwave theory in all aspects of circuit modeling, design, and testing. We need to study the modal field behaviors of the electric and magnetic fields instead of the voltage and current for the high-speed, high-power, and high-density circuits. For this reason, the author presents in this book the EMC modeling and design for high-speed PCBs and advance packaging from the perspective of electromagnetic fields.

In this chapter, we will introduce the major EMC issues related to high-speed PCBs and advanced packaging. Two important passive components will be discussed in detail: the power distribution network (PDN) and through-silicon vias (TSVs) based interposer. We will present SI, PI, and EMI problems of the PDN and interposer, followed by the review of their state-of-the-arts field-circuit hybrid modeling methods. The field–circuit hybrid method is a popular method used for the EMC modeling. It is the bridge that links the obscure electromagnetic equations and the easy-to-understand EMC model. Finally, some practical designs for a high-quality signal propagation and low noise are presented.

1.1 EMC Challenges

Figure 1.1 shows a schematic diagram of a high-speed circuit. Different kinds of chips are heavily mounted on the multilayered PCBs. The circuits are connected together through the vias, bonding wires, bumps/pins, and interconnectors in packages and PCBs. Owing to the increased clock frequency, high density of the devices, and high power consumption, the electromagnetic environment inside the package and PCB is very complex. This results in lots of potential EMC problems, such as the following:

1. *Interconnector delay and loss.* For the increased clock frequency, the length of the interconnector is comparable with the working wavelength. The interconnector must be taken as the transmission line, of which the propagation delay cannot be ignored. At the same time, at a high frequency, the skin effect of the current and the dielectric loss greatly increase. The interconnector loss becomes serious.

2. *Impedance mismatching.* At a high frequency, the impedance mismatching between the interconnectors and the circuits and the discontinuities along the interconnectors will result in multireflections of the signal. This degenerates the signal propagation quality. One of the common interconnector discontinuities in PCBs is the through-hole via. The through-hole via provides electric connection between traces at different PCB layers. However, its parasitic capacitance and residual stub also make the impedance of the interconnector discontinuous.

3. *Simultaneous switching noise (SSN)*. When lots of digital circuits switch at the same time, a heavy current will be drawn from the power supply. Owing to the parasitic inductance and resistance of the power supply system, there will be fast swing of the supplied voltage. This will result in the unstable operation of the integrated circuit (IC). Things become worse for modern high-speed circuits due to the reduced voltage supply level and noise margin.

4. *Cross talk (XT)*. For the dense layout of the circuits and interconnectors, there is strong electromagnetic coupling between them. The signal transmitted on one circuit or interconnector will create interference in another circuit or interconnector. This cross talk increases the bit error ratio of the circuit system.

5. *Unintended antennas*. When the dimensions of the structures in PCBs, such as the traces and the slot on the power–ground planes, are comparable with the working wavelength, they will become effective antennas. Their radiated electromagnetic field will disturb the normal work of the nearby circuit system. One of such unintended antennas is the heat sink on the PCBs.

6. *Susceptibility or immunity*. Above external EMI will induce voltage or current in the IC. This requires that the IC should have certain immunity to protect itself from the external interference or the susceptibility of the IC should be known to make sure that it survives in a complex PCB environment.

There are more EMC problems related to the high-speed circuits than what has been listed earlier. For such a wide variety of EMC problems, in order to establish an easy-to-understand EMC theory system and provide the general international

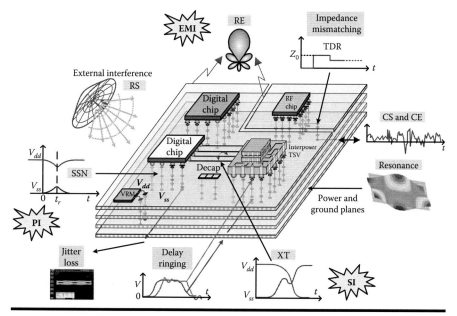

Figure 1.1   EMC problems related to the high-speed circuit.

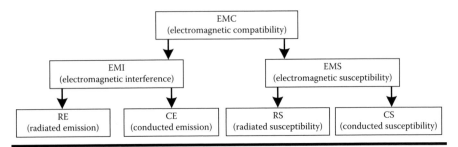

Figure 1.2 EMC problems divided into RE, CE, RS, and CS.

EMC testing standards, EMC problems are categorized according to their common characteristics. On the basis of the source of electromagnetic noise, EMC problems are classified into EMI, where the noise is generated by the circuits and released into the environment, and the electromagnetic susceptibility (EMS), where the noise is from the environment and is coupled to the circuits. On the basis of the noise-coupling path, EMC problems are also classified into radiated mode, where the noise is coupled to the victim through the electromagnetic radiation, and the conducted mode, where the noise is coupled to the victim through the conductors (waveguides or interconnectors). Therefore, EMC problems can be divided into the radiated emission (RE) (such as the unintended antenna), conducted emission (CE) (such as the impedance mismatching and SSN), radiated susceptibility (RS), and conducted susceptibility (CS), as shown in Figure 1.2.

The signal distribution network (SDN) and PDN are necessary structures that compose all high-speed circuits. The EMC problems related to the high-speed circuits are also classified into the SI and PI problems, according to the SDN and the PDN, respectively. Signal integrity is a set of measures of the quality of an electrical signal. It refers to all the EMC problems that arise in high-speed products due to the interconnections [3]. Some of the main SI problems include ringing, cross talk, SSN, impedance mismatching, and signal delay and loss. The PI is the measure of the power supply for the normal work of the high-speed circuits. It is about how to design the PDN to deliver a constant voltage to every IC.

For the high-speed circuit, SI, PI, and EMI (especially the RE) are the three major EMC issues. In the following subsection, some typical SI, PI, and EMI issues related to the high-speed circuit are discussed.

1.1.1 Power Distribution Network

All ICs inside the package need power supply. Owing to the large size of the power device, it cannot be directly connected to the ICs. Therefore, the PDN is employed to deliver voltage from the power device to all ICs. The PDN of the high-speed circuit is composed of a huge number of traces with a very complex layout that have parasitic inductance and resistance. Owing to these parasitic components, there

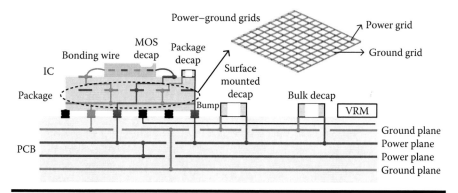

Figure 1.3 **PDNs in package and PCB (cross section).**

will be a voltage drop and ripple along the PDN during the delivery of voltage. At the same time, the PDN is the largest set of conductors inside the package and on the PCB and, therefore, the largest unintended antenna for RE and RS problems.

An entire PDN of the high-speed circuit includes the PDNs in package and PCB. Figure 1.3 shows a schematic diagram of these PDNs, which include the voltage regulator module (VRM), decoupling capacitors (Decaps), power–ground planes (PGPs) or power–ground grids (PGGs), and metal traces connecting them together.

1.1.1.1 Decoupling Capacitors

The VRM is used to change the DC voltage level. It can control the output voltage level by changing the output current. Decoupling capacitors are connected between the power and ground of the PDN and are usually close to the circuits/ICs. They are used to decouple the VRM and the circuits:

1. They work as the circuits/ICs local spare battery and provide the current immediately through their discharging when circuits need a large working current, so as to release the current burden of the VRM. After this, Decaps will be charged by VRM again and will be ready for the next discharging. In this way, the VRM and circuits are decoupled by Decaps.
2. The high-speed circuit generates a high-frequency noise, which couples to and propagates along the PDN. Decaps provide a low-impedance or a short-circuit path for this noise and let it bypass the VRM. Decaps also eliminate the noise propagation inside the PDN.

Decaps in PCB, package, and IC work at different frequency bands [4]. As shown in Figure 1.3, the bulk Decap near the VRM provides the bypass function at low frequencies (several kilohertz to megahertz). The surface-mounted Decap on the

PCB is closer to the chip and can provide a low-impedance path at the middle frequencies (10–100 MHz). For higher frequencies (greater than 100 MHz), Decaps inside the package and ICs are used. For the package and ICs where the silicon is used as the substrate, Decaps are always made of the metal–oxide–semiconductor (MOS) structure.

All kinds of Decaps always have the parasitic inductance and resistance. The parasitic inductance prevents the Decap from providing the current immediately and also increases the Decap impedance at a high frequency. Owing to the parasitic inductance, the application of the bulk and surface-mounted Decap is limited to frequencies less than hundreds of megahertz. In order to increase the working frequency of the Decap, the thin film with a high dielectric constant is embedded in the PCB or package substrate to serve as the Decap. This embedded Decap has a very small parasitic inductance, so it can work at a higher frequency. However, this also increases the fabrication cost. Chapter 5 presents a detailed discussion on Decap and embedded materials.

1.1.1.2 Power–Ground Planes and Power–Ground Grids

When PDN provides instant currents for fast-switching circuits, the spectrum of these currents covers very high frequencies. Owing to the skin effect, current flows mainly on the surface of the conductor at high frequencies. Since the effective cross section of the conductor is reduced, the skin effect causes the effective resistance of the conductor to increase at higher frequencies. In order to reduce the effective resistance of the PDN, usually metal planes, shown in Figure 1.3, instead of metal lines, are employed to deliver the current to circuits, since the metal planes have a larger surface area than the metal lines. At the same time, a complete metal plane inside multilayered PCBs also provides a good electromagnetic shielding between different PCB layers and greatly reduces the noise coupling. However, those metal planes cannot be employed for the package PDN. For the package with the silicon as the substrate, a large metal plane is easy to peel off from the silicon during the fabrication process, and the metal will also be diffused into the silicon. In most of silicon processes, the maximum metal patch size is in the order of several tens of micrometers, so that the PGGs, shown in Figure 1.3, instead of PGPs, are used for package PDN. Such a grid PDN is composed of two metal layers. On each layer, there is alternative distribution of power and ground lines. Through vias are used to connect the power (or ground) lines from the top to the bottom layer to form the power (or ground) network.

Although the PGPs and PGGs can reduce the resistance of the PDN at high frequencies, they also have PI, SI, and EMI problems due to the transmission line effect. In the following, the PGP is used to demonstrate such EMC problems. For the PGG, it has the similar problems. Figure 1.4a demonstrates the PI issue of a power–ground pair, where a circuit is connected to it at the connection point. When the distance between the power supply and the connection point is

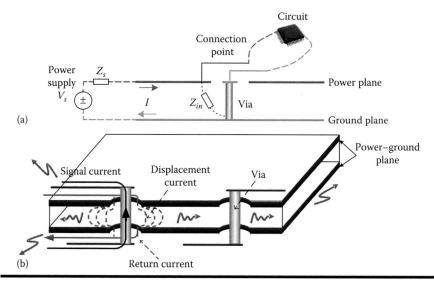

Figure 1.4 EMC problems related to PGPs: (a) PI (cross section) and (b) SI and EMI (From Wei, X. C. and Li, E. P., *IEEE Trans. Microw. Theory Tech.,* **58, 559–565, 2010).**

not smaller enough than the wavelength, owing to the impedance transformation of the transmission line, the transformed internal impedance of the power supply Z_{in} at the connection point is not equal to the original internal impedance of the power supply at high frequencies, $Z_s \neq Z_{in}$. Z_{in}, is also frequency-dependent, and, especially, can be very large at the resonant frequencies of the power–ground pair. This will eliminate the high-frequency components of the supply current and, hence, stop the PDN, providing instant currents with short rising/falling time for fast-switching circuits.

The PGPs also have SI and EMI issues. Inside multilayered PCBs, the horizontal traces are distributed between different power and ground planes, as shown in Figure 1.4b, where each metal plane can be a power or ground plane. Their return currents flow on the PGPs close to them. When the traces pass through different planes, there will be so-called return-path discontinuities, as shown in Figure 1.4b. Therefore, a vertical displacement current is induced between different planes to ensure the continuity of the return currents. These displacement currents will induce electromagnetic noise, which then couples to other signal traces passing through the same PGPs and increases their cross talk. At the same time, this noise also leaks to the surrounding area. In this case, the power–ground pair works as a patch antenna, and the displacement current works as its excitation current. Above interferences will be pronounced if the noise spectrum covers any inherent resonant frequency of the cavity-like pair of PGPs.

1.1.2 Through-Silicon Via

1.1.2.1 3D Integration and Through-Silicon Vias

Moore's law states that the number of transistors in a microprocessor chip will double every two years or so. However, in recent years, when the dimensions shrink to deep submicron for the lithography process, quantum effect and short-channel effect become serious problems. The semiconductor technology gradually achieves its physical limits, and this becomes the bottleneck of Moore's law. A crucial observation from product data in recent years is that transistor density in actual products has not scaled, as would have been expected according to Moore's law. As the principle that has powered the information-technology revolution since the 1960s, Moore's law is nearing its end [6].

For the future development of the semiconductor industry, ITRS lays out a research and development plan on more-than-Moore strategy. Various packages have been developed toward the high-density integration. In particular, a three-dimensional (3D) structure with a TSV technology has emerged as a viable solution for more-than-Moore strategy [7,8]. Unlike the traditional two-dimensional (2D) layouts of the devices inside a chip, 3D IC integration is characterized as a system-level architecture, in which multiple layers of planar devices are stacked and interconnected using TSVs in the vertical direction. A TSV is a coated metal via residing in a silicon substrate for vertical interconnection between stacked dies. The transform from 2D IC integration to 3D IC integration is just like that in a modern city: individual houses are replaced by high-rise buildings in order to achieve more living space. The TSVs, just like the lifts used in those high-rise building, provide fast and efficient electrical connections between different layers of 3D ICs.

Conventional 2D architectures for processors have their L2 cache on a separate die. Processor-to-cache interconnections thus consist of long lines, which slow down the data transfer speed and introduce significant Ohmic loss (as high as 50% of the electrical power is consumed in interconnects). The TSV technology offers a die-to-die stacking, which greatly reduces the interconnect length and increases the bandwidth. Therefore, the TSV technology is probably the best choice for logic and memory applications. In the future, the 3D architecture will integrate different components into one package, such as antennas, sensors, and power management and storage devices. This is known as the heterogeneous integration, as shown in Figure 1.5. It will provide customers with more portable, reliable, and powerful electronic products.

The 3D integration consists of 3D IC packaging, 3D IC integration, and 3D silicon integration [9]. Among all those 3D integrations, TSV-based interposer (passive and active) is preferred by industries due to its easy fabrication and good heat dissipation. Figure 1.6 shows a typical TSV-based interposer. The interposer is a silicon substrate inserted between the die stack and the second-level package. It serves as a space transformer through redistribution by connecting the fine-pitch microbumps to the coarser-pitch C4 bumps [1]. Redistribution layer (RDL) is the

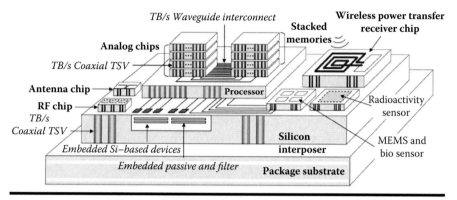

Figure 1.5 Heterogeneous integration (From Professor Joungho Kim at KAIST, Korea).

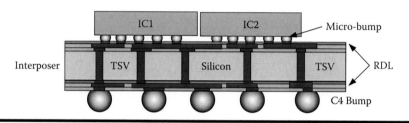

Figure 1.6 A typical TSV-based interposer.

metal layer on the top and bottom of the interposer. It is used for the horizontal interconnection, and TSV is used for the vertical interconnection.

1.1.2.2 EMC Problems Related to TSV

The TSV technology shows many advantages such as shorter interconnect length, simpler exchange interface, greater integration density, and lower power consumption. However, 3D IC and TSVs also introduce new challenges. These include the high density and high aspect ratio via etching; low temperature process for passivation and metallization; high-speed and high-aspect-ratio via filling; thinned wafer/device handling; high-speed and precise die/wafer level alignment and assembly processes (die to wafer and wafer to wafer), testing, and methodology; and competitive cost.

Owing to the reduced size, heat dissipation is one of the most serious challenges in 3D IC designs, and this can degrade the SI. The TSV that can provide vertical heat dissipation through stacked silicon chips is proposed. The TSV can be used to reduce the circuit temperature to a satisfactory level in a 3D IC, which is called thermal TSV [10,11]. Therefore, the thermal TSV insertion cooling scheme is a hot research topic emerging as an effective solution to solve thermal issues in a 3D IC.

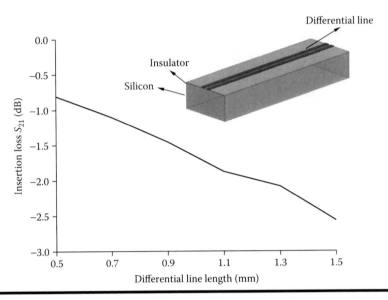

Figure 1.7   Insertion loss of the differential line on the silicon substrate.

From the aspect of the electrical performance of TSVs, we need to consider the loss of silicon interposer, TSVs parasitic inductance, and capacitance effects on high-speed signaling, impedance mismatching, cross talk within dense TSVs arrays, die-to-die vertical coupling, electromagnetic radiation from TSVs, and so on.

Unlike the other substrate materials, silicon is a loss material. Therefore, the signal propagation along horizontal traces and TSV in a silicon substrate will go through a larger IL. Figure 1.7 shows the simulated IL (S_{21}) of a differential line on the silicon substrate, where the conductivity of the silicon is 50 *S/m* and the frequency is 10 GHz. Its decay ratio is about 1.7 dB/mm, which is much larger than that of the same differential line on the FR4 substrate. This loss will increase the bit error rate. For industrial applications, one major EMC concern of the 3D IC and TSV is the loss of a silicon substrate at a high frequency. This frequency-dependent loss also results in signal distortion in the time domain. In [12], an equalization method is proposed to get a flattened IL of the TSV. To reduce the material loss, a couple of glass companies have reported large, thin, and low-cost glass wafers with high quality and their usage for through glass via (TGV) [13]. In comparison with silicon interposer, glass interposer has a high electrical insulation, which is helpful to reduce the signal propagation loss. However, the glass interposer also shows excessive jitter and reduced eye height, as compared with the silicon interposer [14].

Besides silicon loss, there are other EMC risks for the TSV-based interposer. The parasitic inductance of the ground/power TSVs will increase the total impedance of PDN and result in PI problems. The 3D stacked chips also introduce vertical noise coupling due to the short distance between different chips [15]. With the increase in the number of I/Os of highly integrated systems, highly dense integration of TSVs and active circuits on a die has to be realized for a small form factor. Therefore, active circuit-to-TSV cross talk has to be severely considered for the SI in the 3D IC design [16]; this provides a solution to reduce the cross talk between active circuit and TSV by using the guard ring method.

1.1.3 Signal Delay

Complementary metal–oxide–semiconductor (CMOS) devices have high noise immunity and low static power consumption. These properties make them commonly employed in modern high-speed digital chips. The CMOS inverter is a basic device unit in these chips and includes a positive channel-metal-oxide-semiconductor (PMOS) and an negative channel-metal-oxide-semiconductor (NMOS) connecting both gates and both drains together, as shown in Figure 1.8a. The PMOS transistor presents a low resistance between its source and drain contacts when a low gate voltage is applied and presents a high resistance when a high gate voltage is applied. The NMOS transistor presents a high resistance between its source and drain when a low gate voltage is applied and presents a low resistance when a high gate voltage is applied. The low resistance allows the current to flow through it and can be considered the ON state of the PMOS/NMOS, whereas a high resistance limits the current flowing through it and can be considered as the OFF state of the PMOS/NMOS. Therefore, the CMOS inverter can be taken as a switch, as shown in Figure 1.8b. When input voltage is low ("0" level), the NMOS transistor is OFF and the PMOS transistor is ON, and then, the output is connected to the V_{dd} and the current can flow from the power supply to the output. When input voltage is high ("1" level), the NMOS transistor is ON and the PMOS transistor is OFF, and then, the output is connected to the ground V_{ss}.

Owing to the parasitic inductance of the PDN, SI and PI problems could result when the CMOS inverter is connected to PDN and draws the current in it. Two of their major issues are signal delay (or settling time) and SSN.

The high-speed IC includes the core devices and I/O devices. In order to reduce switching time for the core devices between "0" level and "1" level and reduce their

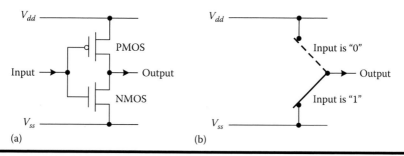

Figure 1.8   **(a) CMOS inverter and (b) its equivalent circuit.**

power consumption, they are usually supplied with a low voltage. Since I/O devices are used to communicate with the circuits outside the IC, their noise level is higher than that of the core devices. In order to increase their noise immunity, I/O devices are supplied with a high voltage. In the following section, the signal delay due to the parasitic inductance of the PDN is briefly introduced for the core and I/O devices.

1.1.3.1 Core Devices

For a core device shown in Figure 1.9a, the output of a CMOS inverter is connected to the gate of the next field-effect transistor (FET). Owing to the MOS configuration of the FET gate, the FET can be taken as the capacitance loading to the previous CMOS inverter. Considering the inverter as a switch, the core device in Figure 1.9a can be equivalent to the RLC circuit of Figure 1.9b, where L is the parasitic inductance of the PDN, C is the capacitance of the FET gate, R means the on resistance of the PMOS of inverter, and $U_s = V_{dd} - V_{ss}$. Initially, the input of the inverter is on "1" level, the FET is connected to the ground, and its gate is on "0" level. At $t = 0$, the input of the inverter changes from "1" to "0" and the FET gate will be charged until it reaches "1" voltage level. This change can be described by using the zero-state series of RLC circuit in Figure 1.9b. For this circuit, its second-order differential equation and initial conditions are

$$LCU_C''(t) + RCU_C'(t) + U_C(t) = U_s, \text{ and} \tag{1.1}$$

$$U_C(0^-) = I_L(0^-) = 0 \tag{1.2}$$

where $U_C(t)$ and $I_L(t) = CU_C'(t)$ are the voltage over C and current flowing through L, respectively.

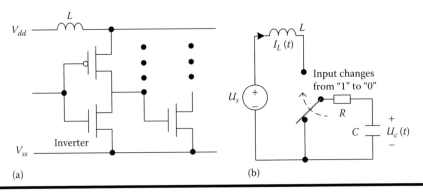

(a)   (b)

Figure 1.9   (a) Core device and (b) its equivalent circuit.

When $(R/2L)^2 > (1/LC)$, the solution of Equation 1.1 is

$$U_C(t) = \frac{U_s s_1 s_2}{s_1 - s_2}\left(\frac{e^{s_1 t}}{s_1} - \frac{e^{s_2 t}}{s_2}\right) + U_s \qquad (1.3)$$

where

$$s_{1,2} = -\frac{R}{2L} \pm \sqrt{\left(\frac{R}{2L}\right)^2 - \frac{1}{LC}} \qquad (1.4)$$

Since $s_{1,2}$ are negative real numbers, magnitudes of $(e^{s_1 t}/s_1)$ and $(e^{s_2 t}/s_2)$ will monotonously decay to zero for $t \to \infty$, and $U_C(t) \approx U_s$ when t is large enough. $U_c(t)$ shows an overdamping response. Therefore, the rising time from "0" to "1" at the FET gate is decided by the decay factors $s_{1,2}$, and they also decide the delay of the signal send from the output of previous inverter and arriving at the input of the next FET.

For $(R/2L)^2 \gg (1/LC)$, from Equation 1.4, we have:

$$s_{1,2} = -\frac{R}{2L} \pm \frac{R}{2L}\sqrt{1 - \frac{1}{LC}\cdot\left(\frac{2L}{R}\right)^2} \approx -\frac{R}{2L} \pm \frac{R}{2L}\left(1 - \frac{1}{2LC}\cdot\left(\frac{2L}{R}\right)^2\right)$$

$$= -\frac{R}{2L} \pm \left(\frac{R}{2L} - \frac{1}{CR}\right) \approx -\frac{R}{L}, -\frac{1}{RC} \qquad (1.5)$$

Equation 1.5 gives two delays: L/R delay and RC delay, which are related to $(e^{s_1 t}/s_1)$ and $(e^{s_2 t}/s_2)$ in Equation 1.3, respectively. Since $(R/2L)^2 \gg (1/LC) \to (R/L) \gg (1/RC)$, the rising time of $U_C(t)$ is mainly decided by the RC delay. Therefore, for a small PDN inductance, its effect on signal delay is not obvious.

When $(R/2L)^2 < (1/LC)$, the solution of Equation 1.1 is

$$U_C(t) = Ke^{-\alpha t}\cos(\omega_d t + \phi) + U_s \qquad (1.6)$$

where $\alpha = (R/2L)$, $\omega_d = \sqrt{(1/LC) - (R/2L)^2}$, $\phi = -\tan^{-1}(\alpha/\omega_d)$, and $K = -(U_s/\cos\phi)$.

From Equation 1.6, we can see that $U_C(t)$ shows an underdamping response. It is an oscillating function with a decay factor α. A smaller L is helpful in reducing the oscillation and getting a shorter settling time.

For the core device with the fixed C and R, the effect of PDN parasitic inductance L on $U_C(t)$ is plotted in Figure 1.10, where $U_s = 1$ V, $C = 1$ nF, $R = 1$ Ω, and $L = 1$ nH and 0.2 nH, respectively. When L is so large that $U_C(t)$ is underdamped, there is overshoot, and it takes a longer time for the signal to achieve its steady state.

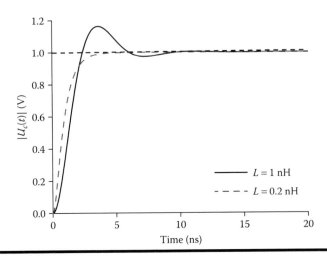

Figure 1.10 Response of $U_c(t)$ under different values of PDN parasitic inductance.

1.1.3.2 I/O Devices

For the I/O device shown in Figure 1.11a, the output of a CMOS inverter is connected to an external interconnector, where the interconnector is located in package or on PCB. Since the length of the interconnector is comparable to the working wavelength, it should be taken as a transmission line with the characteristic impedance Z_0. The equivalent circuit of Figure 1.11a is shown in Figure 1.11b, where Z_L is the load impedance, R means the on resistance of the inverter, and L is the parasitic inductance of the PDN. There are two delays for the signal sent from the output of the inverter and arriving at the load: the L/R delay and the interconnector delay.

In this subsection, it is assumed that $Z_L = Z_0$, so that the load impedance is matched with the characteristic impedance of the interconnector. According to the analysis in subsection 1.1.3.3, under this impedance-matching condition, the interconnector delay is smaller than the L/R delay and is ignored. Therefore, the interconnector is taken as an ideal conductor in circuit analysis. The equivalent

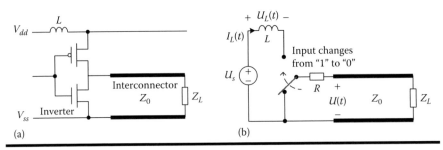

Figure 1.11 (a) I/O device and (b) its equivalent circuit.

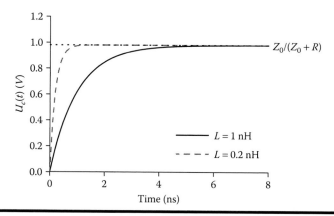

Figure 1.12   Response of U(t) under different values of PDN parasitic inductance.

circuit of Figure 1.11b can be taken as a first-order zero-state series LR circuit, and the voltage at the input of the interconnector is

$$U(t) = \frac{U_s Z_0}{Z_0 + R}\left(1 - e^{-\frac{R}{L}t}\right)$$ (1.7)

$U(t)$ has only the L/R delay, which is decided by the decay factor $-R/L$. Figure 1.12 shows the change of $U(t)$ with t for $L = 1$ nH and 0.2 nH, where $R = 1\ \Omega$, $Z_0 = 50\ \Omega$, and $U_s = 1$ V. For a smaller L, $e^{-(R/L)t}$ approaches to zero quickly, and then, the rising time from "0" to "1" at the interconnector input is faster.

1.1.3.3 Interconnector

It should be noted that $U(t)$ in Figure 1.12 is the signal at the input of the interconnector but not the signal at the load Z_L. For the signal arriving at the load, it will go through another delay related to velocity of the electromagnetic wave propagating along the interconnector. When both load impedance and source internal impedance mismatch with the interconnector characteristic impedance, the interconnector delay or settling time will become large and cannot be ignored. The simplified transmission line model of the interconnector shown in Figure 1.13 is used to demonstrate the effect of such impedance mismatching on the signal delay, where l, Z_0, and v_p are the length, characteristic impedance, and signal propagation velocity of the interconnector, respectively, and R_L is the load resistance. For simplicity, the PDN and CMOS invertors shown in Figure 1.11a are taken as the switch and the equivalent source with a fixed voltage U_s and internal resistance R_s, and the transmission line is taken as lossless.

At $t = 0$, the switch in Figure 1.13 is turned on and the signal voltage will propagate from the source to the load. At $t = T (T = l/v_p)$, the signal arrives at the load.

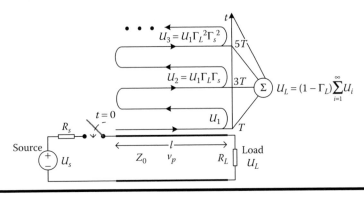

Figure 1.13 Transmission line model of the interconnector.

Considering the impedance mismatching at the load ($R_L \neq Z_0$), part of the signal will be reflected by the load, back to the source. When the reflected voltage arrives at the source at $t = 2T$, due to the impedance mismatching at the source ($R_s \neq Z_0$), it will be partially reflected again by the source forward to the load and arrive at the load at $t = 3T$. This multireflection continues for $t = \infty$. Therefore, the total incident signal voltage at the load can be taken as a summation of all incident voltages arriving at the load at different time, as shown in Figure 1.13.

1. At $t = T$, the first incident voltage arriving at the load is U_1. For $t < T$, there is only incident voltage traveling toward the load, so the transmission line together with the load can be taken as an impedance Z_0. This results in $U_1 = (U_s Z_0 / Z_0 + R_s)$.
2. At $t = 3T$, the second incident voltage arriving at the load is U_2, where $U_2 = U_1 \Gamma_L \Gamma_s$ is the voltage reflected by the load and source. The voltage reflection coefficients of the load and source are defined as

$$\Gamma_L = \left. \frac{\text{reflected voltage}}{\text{incident voltage}} \right|_{\text{at the load}} = \frac{R_L - Z_0}{R_L + Z_0}, \text{ and} \tag{1.8}$$

$$\Gamma_s = \left. \frac{\text{reflected voltage}}{\text{incident voltage}} \right|_{\text{at the source}} = \frac{R_s - Z_0}{R_s + Z_0} \tag{1.9}$$

We have $|\Gamma_L| \leq 1$ and $|\Gamma_s| \leq 1$.
3. At $t = 5T$, the third incident voltage $U_3 = U_2 \Gamma_L \Gamma_s = U_1 \Gamma_L^2 \Gamma_s^2$ arrives at the load.
4. At $t = (2n - 1)T$ for $n = 1, 2, \ldots$, the nth incident voltage $U_n = U_1 \Gamma_L^{n-1} \Gamma_s^{n-1}$ arrives at the load.
5. The total incident voltage at $t = (2n - 1)T$ is

$$U_{total}^{inc}(t)\Big|_{t=(2n-1)T} = \sum_{i=1}^{n} U_i = U_1 \sum_{i=1}^{n}\left[1+\Gamma_L\Gamma_s+\cdots+\Gamma_L^{(n-1)}\Gamma_s^{(n-1)}\right] \qquad (1.10)$$

Owing to the reflection at the load, the total voltage on the load is the total incident voltage U_{total}^{inc} plus its reflected voltage $\Gamma_L U_{total}^{inc}$, which is

$$U_L(t)\Big|_{t=(2n-1)T} = (1+\Gamma_L)U_{total}^{inc}(t)\Big|_{t=(2n-1)T}$$

$$= U_1(1+\Gamma_L)\sum_{i=1}^{n}\left[1+\Gamma_L\Gamma_s+\cdots+\Gamma_L^{(n-1)}\Gamma_s^{(n-1)}\right] \qquad (1.11)$$

Substituting Equations 1.8 and 1.9 into Equation 1.11 and considering the sum formula of the geometric sequence, we have:

$$U_L(t)\Big|_{t=(2n-1)T} = U_1(1+\Gamma_L)\frac{1-\Gamma_L^n\Gamma_s^n}{1-\Gamma_L\Gamma_s} = U_L(\infty)(1-\alpha^n) \qquad (1.12)$$

where $\alpha = \Gamma_L\Gamma_s$, and $U_L(\infty) = (U_sR_L/R_L+R_s)$ is the steady voltage on the load.

From Equation 1.12, we can get the following conclusions:

1. $|\Gamma_L|,|\Gamma_s| \leq 1 \rightarrow |\alpha| \leq 1$, so when t is large enough, U_L is just the voltage across the load when the interconnector is considered as lossless and there is no delay conductor in DC analysis.
2. When either the load or source is matched ($|\Gamma_L| = 0$ or $|\Gamma_s| = 0$), $\alpha = 0$ and $U_L(t)\Big|_{t=(2n-1)T} = U_L(\infty)$. U_L will achieve its steady state after $t > T$. This means that under impedance matching, the interconnector delay is decided only by the time of flight $T = l/v_p$ along the interconnector. This delay could be smaller than other delays, such as the RC delay for core devices and L/R delay for I/O devices, when the interconnector inside the package or on the PCB is short.
3. When both load and source are not matched, the interconnector delay will become large. In the following section, the voltage waveform over the load is analyzed for $\alpha > 0$ and $\alpha < 0$. Let us consider a PCB trace as an interconnector. This PCB trace is a microwave strip line with the length $l = 3\,cm$. It is mounted on the FR4 substrate with relative permittivity of 4.4. Therefore, $T = (l/v_p) \approx (0.03/3\times10^8)\sqrt{4.4} \approx 0.2\,ns$. Assuming $R_s = 1\,\Omega$, $Z_0 = 50\,\Omega$, and $U_s = 1$ V in Figure 1.13, the change of $U_L(t)$ with time under different values of R_L is plotted in Figure 1.14. For $\alpha > 0$, $U_L(t)$ shows an overdamped response, as shown in Figure 1.14a. The larger the load impedance mismatching, the larger the interconnector delay. For $\alpha < 0$, $U_L(t)$ shows an

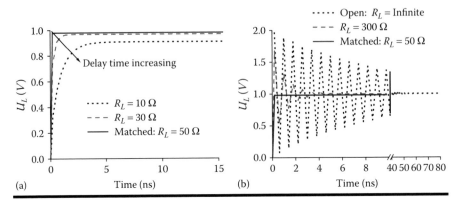

Figure 1.14 **Change of $U_L(t)$ with time, under different values of R_L. (a) $\alpha > 0$ and (b) $\alpha < 0$.**

underdamped response, as shown in Figure 1.14b. The larger the load impedance mismatching, the longer the settling time of $U_L(t)$. When the end of the trace is opened ($R_L = \infty$), the oscillation is so strong that it takes $U_L(t)$ more than 50 *ns* to settle down, in comparison with 0.2 *ns* delay time when the load is matched.

1.1.4 Simultaneous Switching Noise

The voltage level provided by the PND should keep constant for the normal work of circuits. However, when lots of CMOS inverters, shown in Figure 1.9, switch at the same time, a heavy current is drawn from the PDN. Owing to the parasitic inductance and resistance of the PDN, the supplied voltage of the PDN will drop. This will results in a gate voltage of other CMOS inverters and FETs connected to the same PDN drop at the same time, causing unstable operation of a logic gate. This phenomenon is named as simultaneous switching noise. It is a major PI problem for high-density and high-speed circuits. The SSN is also known as the ground bounce, power bounce, and Delta-I noise.

Figure 1.15a shows two CMOS inverters connected to the same PDN. Their equivalent circuits are shown in Figure 1.15b, where $U_s = V_{dd} - V_{ss}$; R_1, R_2, C_1, C_2, and L have the same meanings as those in Figure 1.9b. For inverter 2, its NMOS transistor is OFF and PMOS transistor is ON, and voltage across its load is U_s ("1" level). When the input of the inverter 1 changes from "1" to "0," the current will be drawn from the PDN to change C_1, until it reaches "1" voltage level. By using the voltage $U_{c1}(t)$ over C_1 obtained from Equations 1.3 and 1.6, the voltage across the PDN parasitic inductor L can be calculated as $U_L(t) = Li'(t) = LC_1U_{c1}''(t)$. Therefore, the voltage applied on C_2 will be reduced to $U_{c2}(t) \approx U_s - U_L(t) = U_s - LC_1U_{c1}''(t)$. When $U_{C2}(t)$ is below the threshold, the output of the inverter 2 will be taken as "0" level, which will results in malfunction of the logic gate.

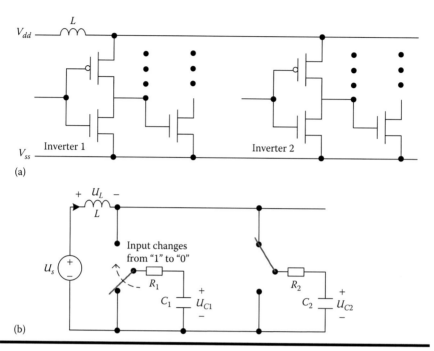

Figure 1.15 (a) Two CMOS inverters connected to the same PDN and (b) their equivalent circuits.

The example in section 1.1.3.1 is used here again to demonstrate the effect of L on $U_{C2}(t)$, where $U_s = 1$ V, $C_1 = 1$ nF, $R_1 = 1\ \Omega$, and $L = 1$ nH and 0.2 nH, respectively. The result is plotted in Figure 1.16. From this figure, we can see that a larger PDN parasitic inductance results in a larger supply voltage fluctuation.

Above-mentioned analysis is based on the simplified CMOS inverter. For practical CMOS, it also shows signal delay. Therefore, U_s in Figure 1.15b should be a step function with a rising time t_r. Figure 1.16 is obtained when the CMOS delay is much smaller than the L/R delay and the RC delay.

When the CMOS delay is larger than the RC delay and the RC delay is much larger than the L/R delay, that is, $t_r \gg RC \gg (L/R)$, the voltage across L will get its maximum value at $t = t_r$, as shown in Figure 1.17a. In Figure 1.17a, the PDN parasitic resistance is considered, so $U_{L\max} \neq U_s$. For this case, the voltage fluctuation over C_2 is shown in Figure 1.17b, where the parasitic inductances of both power lines and ground lines are taken into account. Under $t_r \gg RC \gg (L/R)$, for the core devices shown in Figure 1.15b, $U_{L\max}$ can be approximated as [4]:

$$U_{L\max} = Li'(t)_{\max} \approx \frac{LU_s}{Rt_r} \tag{1.13}$$

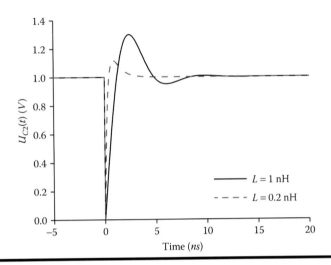

Figure 1.16 Change of $U_{C2}(t)$ with time, under different values of L.

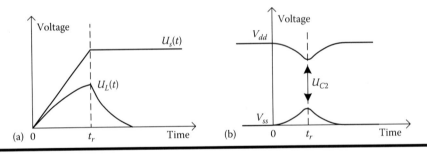

Figure 1.17 (a) Voltage across L and (b) voltage drop across C_2.

where R includes the parasitic resistance of the PDN and the on resistance of CMOS inverter. For the I/O devices in Figure 1.11a, with $t_r \gg (L/R)$, similarly, we have

$$U_{L\,max} = Li'(t)_{max} \approx \frac{LU_s}{(Z_0 + R)t_r} \tag{1.14}$$

Equations 1.13 and 1.14 are useful to estimate the maximum PDN parasitic inductance L for an allowable supply voltage ripple U_{Lmax}/U_s, CMOS delay presented by t_r, and resistance R.

1.1.5 Cross talk

Cross talk (XT) means that a signal transmitted in one circuit or channel creates an undesired effect in another circuit or channel. Owing to the increased working frequency, cross talk has become a serious problem for high-speed circuits and a

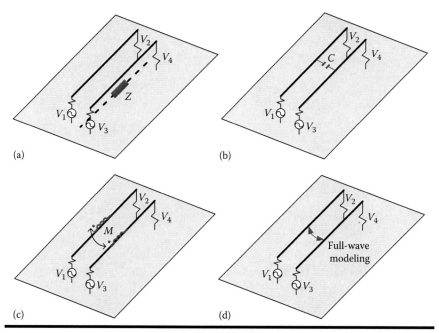

Figure 1.18 Cross talk caused by (a) conductive, (b) capacitive, (c) inductive, and (d) electromagnetic field coupling.

major obstacle of interconnectors' layout. When the circuit size is smaller than the wavelength of interest, the cross talk can be explained by the undesired conductive, capacitive, or inductive coupling from one circuit to another, as shown in Figure 1.18a–c, respectively [17].

1. Conductive cross talk occurs when the aggressor and victim circuits share the same ground. The finite impedance Z of the shared ground results in a voltage drop that appears across both circuits.
2. Capacitive cross talk is due to the parasitic mutual capacitance C between the aggressor and victim circuits; this results in a current leakage from one to another. Usually, capacitive coupling induces a current in the victim circuit that is proportional to the time derivative of the source signal [i.e., $C(dV/dt)$].
3. Inductive cross talk is caused by the parasitic mutual inductance M between the aggressor and victim circuits, which is similar to the coupling between the primary and secondary windings of a transformer.

Above-mentioned parasitic parameters Z, C, and L in Figure 1.18 can be extracted by using the quasi-static software, such as the ANSYS Q3D. When the circuit size is comparable with the wavelength of interest, both electric and magnetic fields have contributions to the coupling between the aggressor and victim circuits,

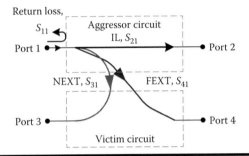

Figure 1.19   Near-end cross talk and far-end cross talk.

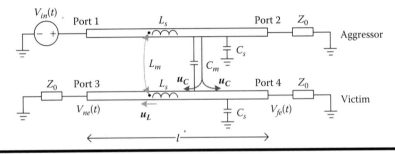

Figure 1.20   NEXT and FEXT voltages of a pair of traces.

as shown in Figure 1.18d. In that case, a single parasitic parameter Z, C, or L is not enough to accurately describe the cross talk. Full-wave simulation must be used to calculate the cross talk.

In EMC engineering applications, the cross talk is classified as near end cross-talk (NEXT) and far end cross talk (FEXT). For the aggressor and victim circuits shown in Figure 1.19, NEXT is the coupling at the same ends of the circuits, whereas FEXT means the coupling at the different ends of the circuits. The scattering parameters are commonly used in EMC measurement. Assuming the signal as input from Port 1 of the aggressor circuit, NEXT and FEXT in Figure 1.19 can be defined by S_{31} and S_{41}, respectively. At the same time, S_{11} and S_{21} are defined as the return loss and the insertion loss (IL), respectively.

For a pair of identical traces, as shown in Figure 1.20, C_s, C_m, L_s, and L_m are the self-capacitance, mutual capacitance, self-inductance, and mutual inductance per unit length of the trace, respectively. Port 1 of the aggressive trace is excited by a pulsed voltage source $V_{in}(t)$, and ports 2 to 4 are terminated with the characteristic impedance Z_0 of the trace. $V_{in}(t)$ will induce NEXT and FEXT on the victim trace through the mutual capacitance C_m and inductance L_m. Assuming that the signal traveling along the aggressive trace is not affected by the victim trace, both capacitively and inductively induced voltages on the victim are proportional to the slope of $V_{in}(t)$. The capacitively coupled voltage u_c on victim will propagate to ports 3 and 4

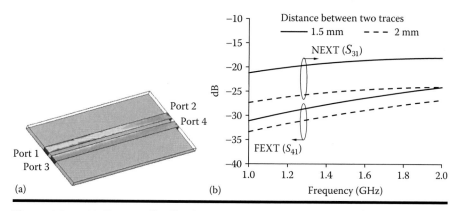

Figure 1.21 **(a) Current distribution on two signal traces at 2 GHz, where port 1 is excited and (b) near-end cross talk and far-end cross talk.**

with same polarity, whereas the inductively coupled voltage u_L will propagate to ports 3 and 4 with opposite polarity.

Assuming that the delay time t_f of the trace is greater than the rising time of $V_{in}(t)$, the analytical formula can be obtained to calculate NEXT and FEXT. The pulse width of the NEXT voltage $V_{ne}(t)$ is equal to a round-trip delay of the aggressive trace [18], and

$$V_{ne}(t) = \frac{1}{4}\left(\frac{C_m}{C_s + C_m} + \frac{L_m}{L_s}\right)\left[V_{in}(t) - V_{in}(t - 2t_f)\right] \qquad (1.15)$$

The FEXT voltage $V_{fe}(t)$ is proportional to the length of the trace [18], and

$$V_{fe}(t) = \frac{1}{2}\left(\frac{C_m}{C_s + C_m} - \frac{L_m}{L_s}\right)\frac{l}{v}\frac{dV_{in}(t - t_f)}{dt} \qquad (1.16)$$

where $l/v = t_f$.

More accurate cross talk results can be obtained by using the full-wave simulation. Figure 1.21a,b plot a pair of coupled signal traces and their full-wave simulated NEXT and FEXT, respectively. From Figure 1.21b, we can see that both NEXT and FEXT will decrease with the increased distance between these two traces. The current distribution on these two traces is plotted in Figure 1.21a at 2 GHz, where signal is an input from port 1. The induced current at ports 3 and 4 can be seen in this figure.

1.1.6 Impedance Mismatching

For high-speed circuits, impedance matching along the whole interconnector is very important to avoid signal distortion. As shown in subsection 1.1.3.3, when characteristic impedance of the interconnector is different from that of the load

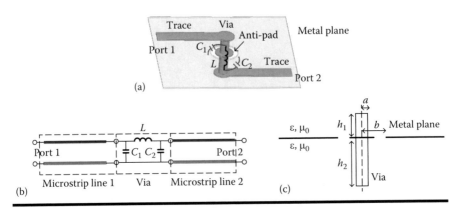

Figure 1.22 (a) Through-hole via, (b) its equivalent circuit, and (c) its dimensions (cross section).

and source, there will be a larger signal delay and a longer settling time. The discontinuities along the interconnectors will also introduce impedance mismatching and result in signal reflection. Such discontinuities could be a slot on the ground plane, a through-hole via, a stub of the through-hole via, and so on.

The through-hole via provides electric connection between traces at different PCB layers; however, its parasitic capacitance and residual stub also make the impedance of the interconnector discontinuous. Figure 1.22a shows a typical through-hole via that goes through the metal plane and connects the traces on the top and bottom layers. The antipad is the air gap between the via and the metal plane. The antipad contributes the parasitic capacitances C_1 and C_2 between the via and the metal plane. The current flowing along the via provides a partial inductance L. Therefore, the via can be equivalent to a PI circuit, as shown in Figure 1.22b. The top and bottom traces in Figure 1.22a can be equivalent to the microstrip lines 1 and 2, respectively, in Figure 1.22b. As shown in Figure 1.22c, the via's PI circuit is inserted between two microstrip lines, which results in the impedance discontinuity between them.

Usually, the parasitic capacitances C_1 and C_2 have a greater effect on impedance mismatching than the parasitic inductance L. Their values can be calculated as [19]:

$$C_{1/2} = \frac{4\pi\varepsilon}{h_{1/2}\ln(b/a)} \sum_{n=1,3,5,\ldots}^{2N-1} \frac{1}{k_n^2}\left[1 - \frac{K_0(k_n b)}{K_0(k_n a)}\right] \tag{1.17}$$

where K_0 is the modified Bessel function of the second kind with zero order. $k_n = \sqrt{(n\pi/2h_{1/2})^2 - (1/\lambda_g^2)}$, with $\lambda_g = (1/f\sqrt{\varepsilon\mu_0})$ being the wavelength in the substrate. Dimensions a, b, and $h_{1/2}$ are shown in Figure 1.22c. If $k_n b > k_n a > 30$, to avoid the computing error for large arguments in Bessel functions, the following asymptotic formulas of Bessel functions are used:

$$C_{1/2} = \frac{4\pi\varepsilon}{h_{1/2}\ln(b/a)} \sum_{n=1,3,5,...}^{2N-1} \frac{1}{k_n^2}\left[1-\sqrt{\frac{a}{b}}e^{k_n(a-b)}\right] \tag{1.18}$$

The partial inductance L can be obtained by considering a finite-length straight conductor in the free space:

$$L = \frac{\mu}{2\pi}\left[h_1\ln\left(0.5413\frac{h_1}{a}\right)+h_2\ln\left(0.5413\frac{h_2}{a}\right)\right] \tag{1.19}$$

The traces + via of Figure 1.22a are simulated to demonstrate the effect of the via on signal propagation, where $a = 0.1$ mm, $b = 0.3$ mm, $h_1 = 0.4$ mm, and $h_2 = 0.5$ mm. FR4 with $\varepsilon_r = 4.4$ and loss tangent = 0.02 is used as the substrate. The total length of the trace is 40 mm. Figure 1.23a,b plot the magnitude and phase of S_{11}, respectively and Figure 1.23c,d plot the magnitude and phase of S_{21}, respectively. The results from the full-wave simulation are used here as a reference. The results from the equivalent circuits of Figure 1.22b with and without (where two microstrip lines are directly connected in Figure 1.22b) the via's PI circuit are compared with result from the full-wave simulation. From Figure 1.23, we can see that the PI equivalent circuit of the via can give an accuracy result in comparison

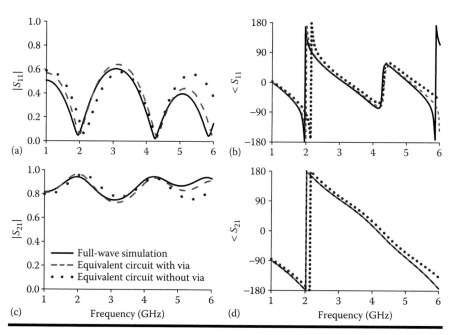

Figure 1.23 (a) Magnitude of S_{11}, (b) phase of S_{11}, (c) magnitude of S_{21}, and (d) phase of S_{21}.

with the full-wave simulation results. A more accurate result can be obtained by considering the parasitic capacitance of the trace bends in the equivalent circuit of Figure 1.22b.

The magnitude fluctuations of S_{11} and S_{21} in three curves of Figure 1.23 result from impedance mismatching between the top and bottom traces (because $h_1 \neq h_2$). By comparing the magnitudes of S_{11} and S_{21} with and without via, we can see that the via changes S_{11} and S_{21} at high frequencies and gives a non-equal amplitude oscillation of S_{11} and S_{21}.

Besides the through-hole via, the slot is another common discontinuous structure in multilayered PCBs. Slots are etched on a metal plane to isolate the noisy circuits from the sensitive circuits or to provide different voltage levels. For a signal trace above the slot, as shown in Figure 1.24, its return current on plane 1 had to find a path to get round the slot: it will flow to plane 2 and back to plane 1 for a slot with a large width, or it will flow from two ends of the slot for a narrow and short slot. These will change the characteristic impedance of the trace part over the slot or introduce additional parasitic inductance and capacitance along the return current path. All of these result in impedance mismatching. The simulated magnitude of S_{21} of the trace, with and without the slot, is plotted in Figure 1.25, where

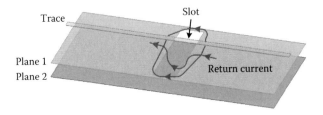

Figure 1.24 Return currents of the trace above a slot.

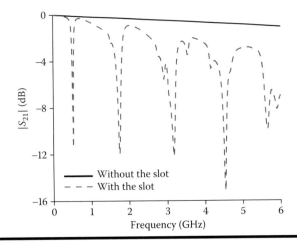

Figure 1.25 Magnitude of S_{21} of the trace, with and without the slot.

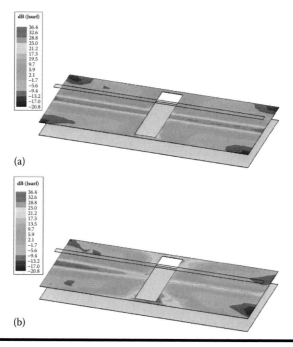

(a)

(b)

Figure 1.26 **Return current distributions on plane 1 at (a) 0.3 GHz and (b) 0.52 GHz (the first resonant frequency).**

the trace is designed with the characteristic impedance of 50 Ω and the width of the slot is 10 mm. From this figure, we can see that the slot reduces the magnitude of S_{21}. Especially, it introduces troughs to S_{21} at some frequencies, which are the resonant frequencies of the slot, and the slot forms a stopband filter for the return current. Figure 1.26a,b plot the return current distribution on plane 1 at a non-resonant frequency (0.4 GHz) and the first resonant frequency (0.52 GHz) of the slot, respectively. From this figure, we can see that at the resonant frequency, the slot greatly eliminates the return current and results in a large reflection.

1.2 EMC Modeling

In the previous section, some typical EMC issues related to high-speed circuits have been discussed. The efficient and accurate modeling and simulation technologies are a great help to better understand those EMC issues and find their solutions. The full-wave methods, such as the finite element method, methods of moments, and finite-difference time-domain, can give an accurate result because they use the mesh to model every detail of the object. Although the overall size of the high-speed circuit is small enough to apply these full-wave methods, the high aspect ratio of the substrate and the huge and tiny structures (narrow slots, through-hole vias, wire bondings, and so on) result in a huge number of mesh. This makes these full-wave methods very expensive in terms of computing time and memory requirement.

In comparison with the full-wave methods, the field–circuit hybrid methods are more efficient. The complex structure is discretized into several parts, and equivalent circuits for every part are extracted individually by solving their electromagnetic fields. By this way, the original complex electromagnetic field problem is simplified to the circuit simulation problem. The circuit simulation is very fast and can explain the inherent mechanism of EMC problems. Therefore, the field-circuit method is preferred by EMC engineers and implemented into much commercial software. In this section, several typical field-circuit methods used to simulate the passive components of the high-speed circuit are discussed.

1.2.1 Field–Circuit Hybrid Method

A complex PCB/Package as shown in Figure 1.1 includes the larger passive components (SDN and PDN) and smaller active devices. Because of the ever-increasing working frequency, the size of the passive components is comparable with the wavelength of interest. The electromagnetic wave effect must be considered, and the passive components should be solved by using electromagnetic field simulators. Usually, the electrically small active devices can usually they can be accurately presented by using the lumped circuit models provided by the venders. The simulation results from passive components, and active devices are then combined to get an efficient solution of the complex PCB/package.

Figure 1.27 shows the typical passive components in a high-speed circuit, including PGPs, interconnectors, and vias. The electromagnetic field can propagate along *x*, *y*, and *z* directions, so it is referred to as the 3D problem. This problem can be divided into following simple two-, one-, and zero-dimensional subproblems [19].

1. Power–ground planes. Because the distance between these metal planes is smaller than the wavelength of interest, it is assumed that the electromagnetic field sandwiched between the pair of planes does not change along the *z* direction. The electromagnetic field propagates only along the *x* and *y* directions, so this plane pair is referred to as a two-dimensional problem. With ports vertically defined between two planes, the plane pair can be equivalent to an *N*-ports network. The network parameters such as [Z] and [S] can be extracted by using a proper two-dimensional method or software, such as the ANSYS SIWave.

2. Signal traces. Since the cross section area of the traces is smaller than the wavelength of interest, the electromagnetic field over the cross section of the traces does not show any wave propagation. Traces support transverse electromagnetic (TEM) mode or quasi-TEM mode, which propagates along them. Therefore, the signal trace is referred to as a one-dimensional problem. The signal traces sandwiched between PGPs are taken as the strip lines, whereas

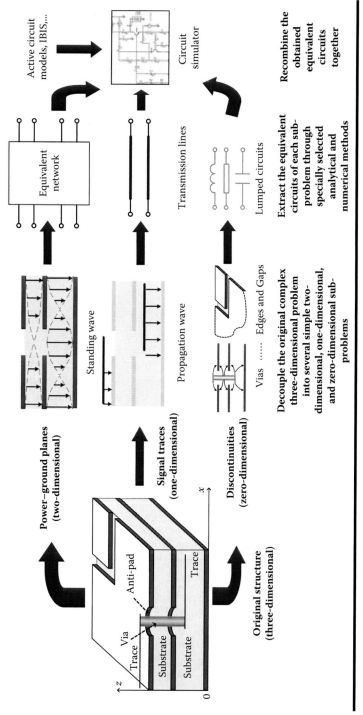

Figure 1.27 Field–circuit hybrid method for the solution of passive components.

the signal traces above or below the PGPs are taken as the microstrip lines. They have different characteristic impedances and propagation constants.

3. Discontinuities such as vias. The dimension of these structures is much smaller than the wavelength of interest, so the electromagnetic field surrounding them is quasi-static and does not show the wave propagation in any direction. In this case, this discontinuity is referred to as a zero-dimensional problem. It can be equivalent to an *RLGC* lumped circuit, and the values of *RLGC* are extracted by using quasi-static methods and software, such as ANSYS Q3D.

Since the original 3D problem is divided into two-, one-, and zero-dimensional subproblems, the number of unknowns and the computing time are greatly reduced. Finally, the above-mentioned extracted equivalent network, transmission line, and lumped circuits are recombined to give the entire equivalent circuit of the original structure. This equivalent circuit can be connected to the circuit models of active devices, and the co-simulation can be performed by using circuit simulators such as the ADS software.

1.2.2 PDN Modeling

The equivalent circuit of PDN shown in Figure 1.3 can be presented by using Figure 1.28 [4]. The Decap is presented by using its capacitance, equivalent series inductance (ESL), and equivalent series resistance (ESR), and PGPs and PGGs are presented by their equivalent network. From this equivalent circuit, we can calculate the input impedance at any node inside the PDN.

The target impedance is a very important indicator for the design of the PDN. The target impedance Z_T is defined as

$$Z_T = \frac{V_{dd} \cdot \text{ripple}}{I_{max}} \tag{1.20}$$

where V_{dd} is the supplied voltage, ripple is the tolerable supplied voltage ripple (usually is 2.5%), and I_{max} is the maximum current drawn by the circuits. For an IC, its power P and supplied voltage are known, so the maximum current drawn by the IC can be estimated as $I_{max} = P/V_{dd}$, according to which the target impedance can be calculated.

If the PDN input impedance shown in Figure 1.28 is larger than the target impedance, then the voltage drop due to the working current flowing through the PDN will be too large for the PDN to provide a stable voltage supply. Currents distribute differently at low and high frequencies to minimize the overall reactance of the PDN [20]. At low frequencies, inductive reactance is negligible, so the currents spread out to minimize resistance. At high frequencies, inductive reactance dominates resistance, so the currents crowd together to minimize inductance, and

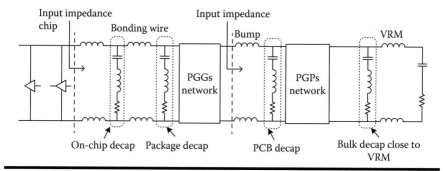

Figure 1.28   Equivalent circuit of the PDN shown in Figure 1.3.

then, resistance is increased. Owing to this redistribution of currents flowing along the PDN with frequencies, PDN impedance is frequency-dependent. Usually, the PDN is designed so that, within the working frequency band, its input impedance is less than the target impedance.

The detailed analysis of the equivalent networks of PGPs and PGGS will be discussed in subsection 1.2.3 and Chapter 2. When the size of PGPs and PGGs is smaller than the interesting wavelength, we can get a simple low-frequency equivalent network. Figure 1.29a shows a power–ground pair in which the plane and substrate

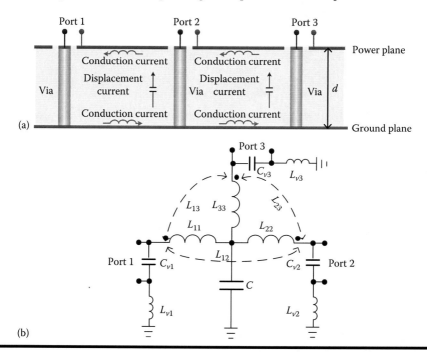

Figure 1.29   (a) Power–ground plane (cross section) and (b) its low-frequency equivalent network.

are assumed to be lossless. Three ports are defined, which can be connected to chips or VRM mounted on the PCB. When the frequency is less than the first-cavity resonant frequency of the power–ground pair, this PGP can be equivalent to an N-ports LC network, as shown in Figure 1.29b. In Figure 1.29b, the parasitic components C and L_{ij} are due to the current distributions between the power and ground planes.

1. C is the capacitance due to the displacement current flowing vertically between the top and bottom planes. $C = \varepsilon S/d$, where ε is the permittivity of the substrate sandwiched between the power and ground planes, S is the area of the PGP, and d is the distance between the power and ground planes. Under low-frequency assumption, C is independent of locations of ports.
2. L_{ij} are the self (for $i = j$) and mutual (for $i \neq j$) inductances. They are used to present the conduction current flowing horizontally on the surfaces of the power and ground planes. Since the value of inductance is dependent on the current patch, for different port locations, $L_{ii} \neq L_{jj}$ with $i \neq j$. The values of L_{ij} can be extracted by using the quasi-static software (such as the ANSYS Q3D) for PGPs with arbitrary shapes.
3. To consider the via effect at every port, its parasitic capacitance and inductance are also included in Figure 1.29b, which are C_{v1}, C_{v2}, C_{v3}, L_{v1}, L_{v2}, and L_{v3}.

It should be noted that the LC equivalent network of Figure 1.29b can also be employed for PGGs. In that case, the value of C should be extracted by using the quasi-static software.

1.2.3 Power–Ground Pair Modeling

At the low-frequency, the power–ground pair (where each metal plane can be power or ground plane) can be equivalent to the lumped circuit, as shown in Figure 1.29b. When the frequency increases, the more accurate 2D methods must be used. Different 2D methods have been extensively employed in the literature.

1. When the power–ground pair has regular shapes, such as rectangles, circles and triangles, a closed form of the Green's functions can be formulated, which results in an impedance formula in terms of the summation of infinite number of resonant modes. This approach will be discussed in detail in Chapter 2.
2. For the power–ground pair with arbitrary shapes, the 2D numerical methods are developed.
   a. The 2D integral equation method, also called the contour integral method, has been used in [19] and [21] to study general PGPs with arbitrary shapes, which will be discussed in Chapter 3.

b. The 2D finite-element method (2D FEM) is also used to simulate PGPs [22]. The 2D FEM method had been implemented in to the commercial software, such as ANSYS SIWave (from ANSYS company).

c. The 2D finite-difference time-domain method (2D FDTD) has also been used to model PGPs [23]. Furthermore, a multilayered finite-difference method was proposed in [24].

In the following subsection, the 2D finite-difference method and a semi-analytical method are discussed for the power–ground pair.

1.2.3.1 2D Finite-Difference Method

Figure 1.30a shows the power–ground pair with an arbitrary shape, where the substrate has a complex electric permittivity of $\tilde{\varepsilon} = \varepsilon(1 - j\tan(\delta))$ and ε and δ are the real electric permittivity and loss angle of the substrate, respectively, with $\tan(\delta) = (\sigma/\omega\varepsilon)$. When the thickness of the substrate d is smaller enough than the wavelength of interest, we have the following 2D wave equation:

$$\left(\frac{\partial^2}{\partial x^2} + \frac{\partial^2}{\partial y^2} + \omega^2 \tilde{\varepsilon} \mu \right) u(x, y) = -j\omega\mu d J_z(x, y) \qquad (1.21)$$

where u is the voltage defined between the two metal planes and J_z is the excitation current density along the z direction. J_z presents the return current density flowing along Z_{in} in Figure 1.4a or the displacement current in Figure 1.4b.

The power–ground pair in Figure 1.30a can be discretized into many small square cells with a side length of h, as shown in Figure 1.30b. The bottom metal

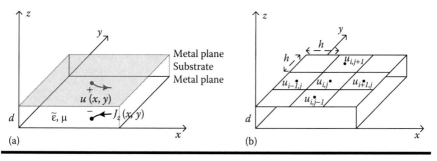

Figure 1.30 (a) The power–ground pair and (b) its discretization.

plane is taken as a reference plane, and $u_{i,j}$ and $J_{z\,i,j}$ present $u(x,y)$ and $J_z(x,y)$ at the centre of (i,j) cell, respectively. By using the following difference formula:

$$f''(x)_{x=0} \approx \frac{f(h)+f(-h)-2f(0)}{h^2} \qquad (1.22)$$

Equation 1.21 is discretized as

$$u_{i+1,j}+u_{i-1,j}+u_{i,j+1}+u_{i,j-1}-4u_{i,j}+h^2\omega^2\tilde{\varepsilon}\mu u_{i,j}=-j\omega\mu dh^2 J_{zi,j} \qquad (1.23)$$

Now, we define the inductance $L=\mu d$ between two adjacent cells, cell's capacitance $C=(\varepsilon h^2/d)$, and conductance $G=(\omega\varepsilon h^2 \tan\delta/d)$. Equation 1.23 can be written as

$$\frac{u_{i+1,j}-u_{i,j}+u_{i-1,j}-u_{i,j}+u_{i,j+1}-u_{i,j}+u_{i,j-1}-u_{i,j}}{j\omega L}=-I_{zi,j}+(j\omega C+G)u_{i,j} \qquad (1.24)$$

Therefore, the equivalent circuit of the unit cell can be obtained as: where $R=(2/\sigma_c t)+\sqrt{(j\omega\mu_c/\sigma_c)}$ is added by considering the losses of the metal planes, and t and σ_c are the thickness and conductivity of the metal planes. $I_{zi,j}=h^2 J_{zi,j}$, and a port$_{ij}$ is defined at the center of (i,j) cell, with $I_{z\,i,j}$ as the port current.

According to Figure 1.31, the whole power–ground pair of Figure 1.30a can be taken as a 2D transmission line, as shown in Figure 1.32, where the port current $I_{z\,i,j}$ can be added, when necessary. It should be noted that in the equivalent circuits of Figure 1.31 and Figure 1.32, since the bottom plane is taken as a reference plane, the inductance and resistance related to the bottom plane are combined with those of the top plane.

The square power–ground pair shown in Figure 1.33a is used to verify the accuracy of the equivalent 2D transmission line, where the side length of the power–ground pair is 62.5 mm. The substrate has a relative permittivity of 4, a thickness of 0.5 mm, and is assumed to be lossless. A port is defined at ANSYS company, of which the input impedance is calculated and plotted in Figure 1.33b. In Figure 1.33b, "equivalent circuit—open" means the result from the equivalent 2D transmission line and "full-wave method" is the result from the full-wave simulation, which is used as the reference. From this figure, we can see a good agreement between them.

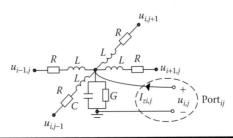

Figure 1.31 The equivalent circuit of unit cell of the power–ground pair.

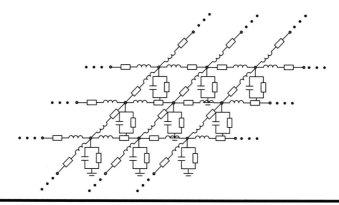

Figure 1.32   The equivalent circuit of the whole power–ground pair.

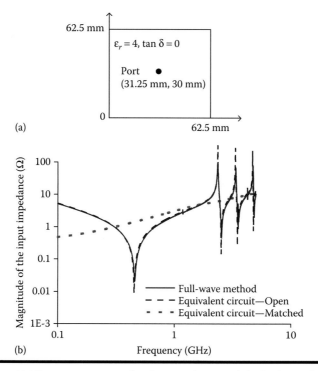

(a)

(b)

Figure 1.33   (a) The power–ground pair (top view) and (b) its input impedance.

For the power–ground pair in the free space, the ends of all transmission lines in Figure 1.32 are taken as open circuits. This boundary condition makes the power–ground pair like a resonator. The peaks of the input impedance in Figure 1.33b result from the resonance of the power–ground pair. If all transmission lines of Figure 1.32 are terminated with their characteristic impedance $\sqrt{(L/C)}$, those peaks will be removed. This is shown in Figure 1.33b by "equivalent circuit—matched."

The authors of [24] had given a detailed discussion about this finite difference method, including the through-hole via coupling and aperture coupling between multilayered planes, penetration through PGPs, and so on.

1.2.3.2 Scattering Matrix Method

Full-wave simulation methods are accurate for the PDN; however, they suffer from long computing times and large memory requirements [21]. This is due to the dense meshing of the small vias inside the power–ground pair. The combination of analytical and full-wave methods will alleviate the computational cost of the full-wave methods. A cylindrical wave-expansion method combined with the Foldy-Lax formula was applied to the analysis of a large number of vias in PCBs [25–29]. In this method, the analytical cylindrical wave expansion is used to model vias, which greatly increases the efficiency of the full-wave method.

1.2.3.2.1 Cylindrical Expansion Combined with Foldy-Lax Formula

A multilayered parallel plate with multiple through-hole vias is used to demonstrate the cylindrical expansion, combined with the Foldy-Lax formula, where the vias are modeled as perfectly conducting cylinders. Figure 1.34 shows one pair of PGPs inside the multilayered parallel plates.

According to the modal theory, the electromagnetic fields around each via can be decomposed into TE (electric field is transverse to z direction) modes and TM (magnetic field is transverse to z direction) modes. Total electromagnetic fields are rewritten according to the components tangential to the $\hat{\rho}$ direction in the ith parallel-plate planes, which comprise the incident and reflected fields of each via:

$$
\begin{aligned}
E_t^{(i)}(\rho,\phi,z) \\
= \sum_{n=-\infty}^{\infty} & \left\{ \sum_{m=0}^{\infty} \left[a_{mn}^{E(i)} J_n(k_m^{(i)}\rho) + b_{mn}^{E(i)} H_n^{(2)}(k_m^{(i)}\rho) \right] e_{tmn}^{E(i)} \right. \\
& \left. + \sum_{m=1}^{\infty} \left[a_{mn}^{H(i)} J_n{}'(k_m^{(i)}\rho) + b_{mn}^{H(i)} H_n'^{(2)}(k_m^{(i)}\rho) \right] e_{tmn}^{H(i)} \right\} e^{jn\phi}
\end{aligned}
\tag{1.25}
$$

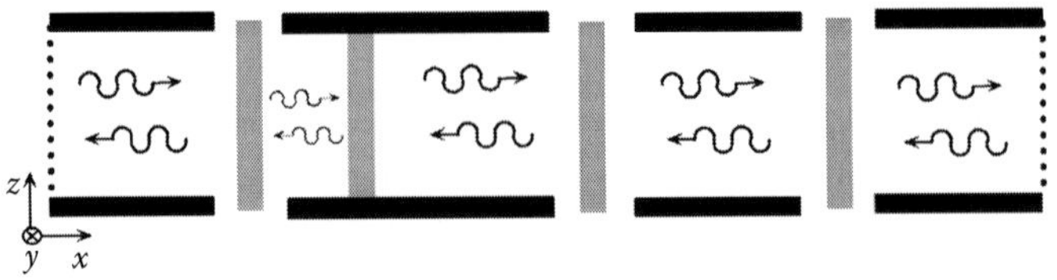

Figure 1.34 A cross-sectional view of one layer of a parallel-plate structure formed by two adjacent conductor planes with many vias.

$$\boldsymbol{H}_t^{(i)}(\rho,\phi,z)$$

$$= \sum_{n=-\infty}^{\infty} \left\{ \sum_{m=0}^{\infty} \left[a_{mn}^{E(i)} J_n{}'(k_m^{(i)}\rho) + b_{mn}^{E(i)} H_n'^{(2)}(k_m^{(i)}\rho) \right] \boldsymbol{b}_{tmn}^{E(i)} \right. \tag{1.26}$$

$$\left. + \sum_{m=1}^{\infty} \left[a_{mn}^{H(i)} J_n(k_m^{(i)}\rho) + b_{mn}^{H(i)} H_n^{(2)}(k_m^{(i)}\rho) \right] \boldsymbol{b}_{tmn}^{H(i)} \right\} e^{jn\phi}$$

where for the *mn*th TM mode,

$$\boldsymbol{e}_{tmn}^{E(i)} = \cos\left(\beta_m^{(i)}\left(z - z_i\right)\right)\hat{z} - \frac{jn\beta_m^{(i)}}{k_m^{(i)2}\rho} \sin\left(\beta_m^{(i)}\left(z - z_i\right)\right)\hat{\phi}$$

$$\boldsymbol{b}_{tmn}^{E(i)} = -\frac{j\omega\varepsilon}{k_m^{(i)}} \cos\left(\beta_m^{(i)}\left(z - z_i\right)\right)\hat{\phi} \tag{1.27}$$

and for the *mn*th TE mode,

$$\boldsymbol{e}_{tmn}^{H(i)} = \frac{j\omega\mu}{k_m^{(i)}} \sin\left(\beta_m^{(i)}\left(z - z_i\right)\right)\phi$$

$$\boldsymbol{b}_{tmn}^{H(i)} = \sin\left(\beta_m^{(i)}\left(z - z_i\right)\right)\hat{z} + \frac{jn\beta_m^{(i)}}{k_m^{(i)2}\rho} \cos\left(\beta_m^{(i)}\left(z - z_i\right)\right)\hat{\phi} \tag{1.28}$$

$\beta_m^{(i)} = m\pi/h_i$, with m being the number of waveguide modes and h_i being the thickness of the ith parallel-plate plane. $z \in (z_i, z_i + h_i)$. The wavenumbers have the relation of $(k_m^{(i)})^2 = k_i^2 - (\beta_m^{(i)})^2$, with $k_i = \omega\sqrt{\mu_i \varepsilon_i}$.

The total electromagnetic fields inside the PGPs are classified as the incident waves and reflection waves of each via. $a_{mn}^{E/H}$ and $b_{mn}^{E/H}$ represent the expansion coefficients of incident waves and reflected waves of each via, respectively. They are the unknowns to be determined. After obtaining their values, the electromagnetic fields at any location inside the PGPs can be calculated by using Equations 1.25–1.28, and consequently, the equivalent network of the PGPs can be obtained. $a_{mn}^{E/H}$ and $b_{mn}^{E/H}$ are dependent on each other. They are related by $b_{mn}^{E/H} = T_{mn}^{E/H} a_{mn}^{E/H}$, with $T_{mn}^{E/H}$ being the reflection coefficient of each via. Therefore, only $b_{mn}^{E/H}$ are required to be solved.

The field expansions must be transferred into linear equations, which can be solved by computer. To do so, the Foldy-Lax equations is used:

$$\boldsymbol{b} = [\boldsymbol{T}] \cdot (\boldsymbol{a} + \boldsymbol{a}') \tag{1.29}$$

where $\boldsymbol{a}$ is the vector representing the external incident wave, $\boldsymbol{b}$ is the vector representing reflected waves away from each cylinder, and $\boldsymbol{a}'$ is the vector representing internal incident waves. a' is produced by multireflection between vias. $[T]$ is a block diagonal matrix, which is an assembly of all the reflection coefficients of each cylinder. The addition theorem for Bessel functions is employed to get $a' = [S]b$, where $[S]$ is the transform matrix, which transfers the reflected waves away from one cylinder into the incident wave coming to another cylinder. Finally, the linear equation is obtained as

$$b = [T] \cdot (a + [S]b) \qquad (1.30)$$

which is solved, and then, the $b_{mn}^{E/H}$ can be obtained.

1.2.3.2.2 Frequency-Dependent Cylinder Layer

Because of the difference between the substrate sandwiched between the power and ground planes and the surrounding air space, there will be wave reflection from the periphery of the PGPs. Usually, the periphery of each PGP pair can be taken as a perfect magnetic conductor (PMC) wall, that is, open-circuited. The above-mentioned modal expansion method has difficulty in modeling such a periphery. This is because the shape of the arbitrary periphery is not natural to cylindrical modes. If the periphery is not accurately modeled, the resonance due to the reflections from the periphery cannot be correctly obtained. The modal expansion method had been used where the periphery of the PGPs is assumed to be circular, in order to make the shape natural to the cylindrical modes. Unfortunately, this result cannot be extended to general cases in electronic package design.

To solve this problem, a novel boundary modeling method, named the frequency-dependent cylinder layer (FDCL), is proposed in [26–28]. The FDCL is devised such that virtual PMC cylinders are postulated to replace the original continuous boundary of each layer in a parallel-plate structure, as shown in Figure 1.35.

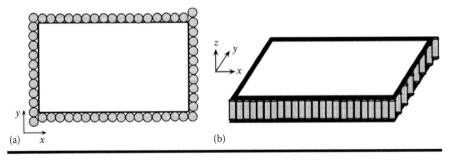

Figure 1.35 An illustration of the implementation of the FDCL boundary, where a series of virtual PMC cylinders (shaded) is placed at the periphery to replace the original continuous finite boundary of a package structure: (a) top view and (b) three-dimensional view.

The PMC cylinders are arranged such that they touch each other and are tangent to the original boundary. Their radii vary with different simulation frequencies.

The advantage of this FDCL is that the PMC cylinders are compatible with the modal expansion with the Foldy-Lax equations; this has demonstrated the significance and efficiency of FDCL in package modeling. The original linear system of equations needs to be only slightly modified and augmented by incorporating the reflection coefficients of the PMC cylinders in the FDCL into Equation 1.30.

The radius of the fictitious PMC cylinder should be carefully chosen. For each pair of PGPs, the dominant mode is the TM mode, where the polarization of the electric field is along z direction and the polarization of the magnetic field is within the xy plane. Therefore, the polarizations of the magnetic field and the PMC cylinders are vertical to each other. In this case, if the radii of the PMC cylinders are too small compared with the operating wavelength, the electromagnetic field will penetrate the fictitious PMC cylinders, and they fail to represent the PMC boundary. On the other hand, if their radii are too large compared with the operating wavelength, the PMC cylinders cannot accurately model the periphery due to the gaps between the PMC cylinders and the real periphery. Therefore, the radii of the PMC cylinders should be changed with the operating frequencies.

An empirical formula to determine the radii of the PMC cylinders in the FDCL is

$$k_g \cdot r_c = \xi \tag{1.31}$$

where $k_g = 2\pi/\lambda_g$, and λ_g is the guide wavelength determined by the electrical properties of the substrate material. ξ is a predefined constant to adjust the frequency-dependent radii of the PMC cylinders in the FDCL. Numerical experiments reveal that an optimal value of ξ is around 0.5 for accurate simulation results.

1.2.3.3 Connection of Power–Ground Pairs

After ports are properly defined, the equivalent network of individual power–ground pair, as shown in Figures 1.29 and 1.32, can be connected with other power–ground pair or signal traces to perform the PI and SI analyses. Figure 1.36a shows two power–ground pairs: PGP1 and PGP2. Port 1 of PGP1 is connected to port 3 of PGP2 through a shorting via. Port 2 of PGP1 is used to provide voltage to other circuits. Figure 1.36b shows the connection of their equivalent networks. L_1, L_2, and L_3 are the vias partial inductances, and C_1, C_2, and C_3 are the capacitances between the vias and planes, the values of which can be extracted from the quasi-static software.

Figure 1.37a plots one power–ground pair with a signal trace passing through it, where two ports are defined at two ends of the signal trace. Figure 1.37b shows their equivalent circuit. C and L are the parasitic capacitance (including the via and the bend) and the inductance of the through-hole via, respectively. The PGP means the equivalent network of the power–ground pair. The signal current and its return

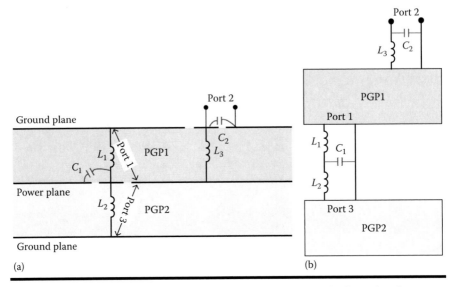

Figure 1.36 (a) Two power–ground pairs and (b) their equivalent circuits.

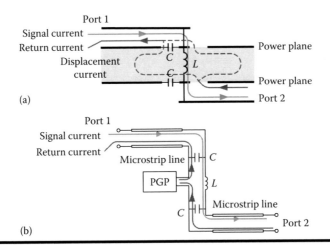

Figure 1.37 (a) One power–ground pair with a signal trace passing through it and (b) its equivalent circuits.

current are plotted in Figure 1.37a,b. The displacement current flows into the PGP. At the resonant frequencies of the PGP, the return current will be stopped, and hence, there will be a large signal reflection. Therefore, by using the equivalent circuit in Figure 1.37b, we can better understand this SI problem caused by the power–ground pair, which shows the advantage of the field–circuit hybrid method over the full-wave method.

1.2.4 Through-Silicon Vias Modeling

Some approaches have been proposed to simulate the electrical performance of the TSV. For single TSV modeling, different-shaped TSVs, including the tapered TSV, rectangular TSV, coaxial TSV, and shielded TSV, are studied. In [30], a modal basis functions is used to describe the polarization current density distribution in the insulator, and the electric field integral equation is employed and solved to extract the equivalent circuits of the TSVs. Other researchers proposed a physical-based equivalent circuit model of dense TSVs, where the values of the *RLCG* in the equivalent circuit are obtained through analytical methods [31]. For multiple die stacking, the cascaded scattering matrix approach discussed in the previous subsection is used [32]. In [33], the TSV–TSV noise coupling is analyzed by using 3D transmission line matrix method. The cross talk within dense TSVs array in a 3D IC is analyzed in [34]. An equivalent circuit is proposed for the analysis of the noise coupling between TSV and active devices in [35].

The TSVs are typically 10 to 100 um long, which is much smaller than the working wavelength in the silicon substrate. Therefore, they can be accurately modeled by using lumped circuits up to several tens of gigahertz. For signal TSV configurations, they can be GS (ground-signal), GSG, GSGSG, and many other layouts. Ground-signal layout is the basic differential line layout for signal TSVs; GSG of TSVs can be connected to the coplanar waveguide (CPW), which has a better shielding performance than the GS layout. Figure 1.38 shows the equivalent lumped circuit model of the GS TSVs. This model can be easily extended to other layout of TSVs. In Figure 1.38, each TSV is composed of the copper core and a thin SiO₂ layer. The SiO₂ layer is used to isolate the copper from the silicon substrate. Owing to this MOS configuration of TSV, when the electric potentials of the TSV and silicon are different, there will be depletion region at the SiO₂–silicon interface.

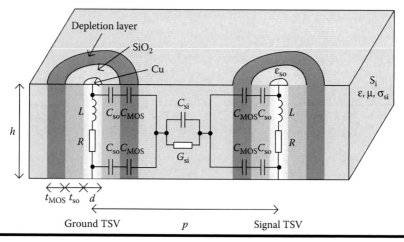

Figure 1.38 A pair of TSVs and its equivalent lumped circuit.

The width of the depletion layer t_{MOS} is decided by the voltage between the TSV and silicon [36]. The GS TSVs are equivalent to the two-conductor transmission line in Figure 1.38. Different analytical formulas are provided in published papers to extract the parasitic parameters. In the following, based on the two-conductor transmission line theory, the simplified formulas are provided.

R is due to the ohm loss of copper. Considering the skin effect of copper at a high frequency, R is written by using the root mean square of its DC and AC values.

$$R = \sqrt{R_{DC}^2 + R_{AC}^2} \tag{1.32}$$

with

$$\begin{cases} R_{DC} = \dfrac{\rho_{cu} h}{\pi r^2} \\[3mm] R_{AC} = \dfrac{k \rho_{cu} h}{\pi (d - 2\delta) \delta} \end{cases} \tag{1.33}$$

where all dimensions are shown in Figure 1.38. $r = d/2$, $\delta = \sqrt{(\rho_{cu}/\pi f \mu)}$ is the skin depth of copper at the frequency f, ρ_{cu} is the resistivity of the copper, and μ is the permeability of vacuum. At a high frequency, the current crowds face-to-face on the ground and signal TSVs. This proximity effect is considered in R_{AC} by the parameter k [37].

L is the parasitic inductance of each TSV, and C_{so} is the capacitance of the SiO$_2$ layer. Taking the TSV as a coaxial structure, with inner diameter d and outer diameter $d + 2t_{so}$, the values of L and C_{so} are calculated as

$$L = \frac{\mu h}{2\pi} \ln \frac{r + t_{so}}{r} \tag{1.34}$$

$$C_{so} = \frac{\pi \varepsilon_{so} h}{\ln(r + t_{so}/r)} \tag{1.35}$$

where ε_{so} is the permittivity of the SiO$_2$.

C_{MOS} is the MOS capacitance of the depletion layer, which is also calculated according to its coaxial structure.

$$C_{MOS} = \frac{\pi \varepsilon_{si} h}{\ln(r + t_{so} + t_{MOS}/r + t_{so})} \tag{1.36}$$

where ε_{si} is the permittivity of silicon.

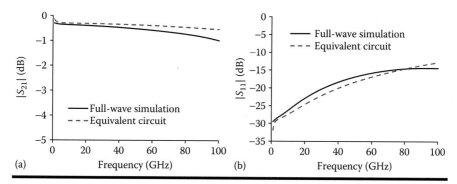

Figure 1.39 Magnitudes of (a) S_{11} and (b) S_{21} for a pair of TSVs.

C_{si} and G_{si} are the capacitance and conductance, respectively, of silicon substrate. Their values can be calculated by considering the two TSVs as two conductors with the distance p and diameter $d_e = d + 2t_{so} + 2t_{MOS}$.

$$C_{Si} = \frac{\pi \varepsilon_{Si} h}{\ln\left((p/d_e) + \sqrt{(p/d_e)^2 - 1} \right)} \tag{1.37}$$

$$G_{Si} = \frac{\pi \sigma_{Si} h}{\ln\left(p/d_e + \sqrt{(p/d_e)^2 - 1} \right)} \tag{1.38}$$

where σ_{si} is the conductivity of silicon.

Figure 1.39a,b show the simulated IL and return loss of the GS TSVs, where the parameters are $h = 100$ um, $r = 20$ um, $p = 150$ um, $t_{so} = 1$ um, and $t_{MOS} = 0$ um. The relative permittivity of SiO_2 and silicon is 4 and 11.9, respectively, and the conductivity of silicon is 10s/m. From this figure, we can see that there is a good correlation between results from the equivalent circuit of Figure 1.38 and the full-wave simulation. However, when the frequency increases to several tens of gigahertz, there is a large error in the equivalent circuit results. At that time, the TSV cannot be presented by a simplified lumped circuit model, and the full-wave method must be used to get a more accurate result.

1.2.5 *Partial Element Equivalent Circuit Method*

Partial element equivalent circuit (PEEC) is a typical field–circuit hybrid method. For the PEEC method, usually the electromagnetic field integral equation is solved to extract the partial inductance, resistance, and capacitance of the structure, so the problem will be transferred from the electromagnetic domain to the circuit domain, where conventional simulation program with integrated circuit emphasis (SPICE)-like circuit simulators can be employed to analyze the equivalent circuit. There are several advantages of the PEEC method over the full-wave methods, such as one can easily integrate any electrical component (*RLCG* components, sources, non-linear elements, ground, etc.) into the PEEC circuit. At the same time, using the PEEC circuit, it is easy to exclude capacitive, inductive, or resistive effects from the model, when it is possible, in order to make the model smaller. For example, in the power electronics application, where the magnetic field is a dominating factor over the electric field, the PEEC circuit can be simplified by just neglecting all capacitance in the model.

The PEEC has become a very popular method of analysis within the field of electromagnetic compatibility, EMI, electrical interconnect problem, and SI. It was first proposed by Albert Ruehli at IBM Thomas J. Watson Research Center [38]. After that, lots of research efforts have been made and many variants of PEEC models have been developed. A clear and easy notation has been devised to differentiate between those different PEEC models. For example, the notations L_p, P, R, and τ of PEEC mean that the model includes the partial inductance L_p, coefficients of potential P, resistance R, and delays τ, respectively. For a specific application, other combinations of elements may be more suitable [39]. Software and tools based on PEEC algorithm had been developed by different companies, such as the StatMod and PCBMod, TPA, and PowerPEEC [40]. The PEEC, together with the FastCap [41] and FastHenry [42], developed at MIT, has become a powerful tool to extract the parasitic parameters in a high-speed circuit.

In the following, the basic idea of the PEEC method in the frequency domain is discussed through the parameters extraction of a one-dimensional signal trace.

1.2.5.1 Electric Field Integral Equation

Figure 1.40 shows one trace immersed in an effective homogeneous medium characterized by the permeability μ and permittivity ε (for the microstrip line-type trace, its effective permittivity is used as ε). I is the current flowing along the trace,

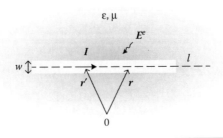

Figure 1.40   A trace, with the current flowing along it.

l is the axis of the trace, and w is the width of the trace. $\boldsymbol{E}^e$ represents the applied electric field, which can be from the external incident wave or from an internal voltage source. On the surface of the trace, the tangential electric field radiated by $\boldsymbol{I}$ plus the tangential component of $\boldsymbol{E}^e$ should be equal to the surface current density multiplied by the surface impedance of the trace, which gives:

$$\left[-j\omega \boldsymbol{A}(\boldsymbol{r})-\nabla\Phi(\boldsymbol{r})+\boldsymbol{E}^e(\boldsymbol{r})\right]_t = Z_s \frac{I(\boldsymbol{r})}{w}, \ \boldsymbol{r}\in l \tag{1.39}$$

where the subscript t means the tangential component. Equation 1.39 is also named as the electric field integral equation.

$Z_s = (1+j)\sqrt{(\pi f \mu/\sigma)}$ is the surface impedance of the trace, f is the frequency, and σ is the conductivity of the trace. The electric field radiated by $\boldsymbol{I}$ is $-j\omega \boldsymbol{A}(\boldsymbol{r})-\nabla\Phi(\boldsymbol{r})$, where the vector potential is

$$\boldsymbol{A}(\boldsymbol{r}) = \mu\int_l G(\boldsymbol{r},\ \boldsymbol{r}')\cdot \boldsymbol{I}(\boldsymbol{r}')dl' \tag{1.40}$$

and the scalar potential is

$$\Phi(\boldsymbol{r}) = \frac{1}{\varepsilon}\int_l G(\boldsymbol{r},\boldsymbol{r}')q(\boldsymbol{r}')dl' \tag{1.41}$$

$G(\boldsymbol{r},\boldsymbol{r}') = (e^{-jk|\boldsymbol{r}-\boldsymbol{r}'|}/4\pi|\boldsymbol{r}-\boldsymbol{r}'|)$ is the free-space Green's function, k is the wave number, and $\boldsymbol{r}$ and $\boldsymbol{r}'$ are the observation and source points, respectively. $q(\boldsymbol{r}')$ is the line electric charge density along the trace. According to the conservation law of electric charge, we have:

$$q(\boldsymbol{r}') = -\frac{1}{j\omega}\frac{\partial I(\boldsymbol{r}')}{\partial l} \tag{1.42}$$

The trace is divided into N small segments, as shown in Figure 1.41, which are named as *current segments*. We use l_i to present the ith current segment and also its length. r_i^+ and r_i^- denote the front and back nodes of the ith current segment, respectively. The current segments are used to expand the current $\boldsymbol{I}$, and $\boldsymbol{I}$ is assumed to be constant on every segment. Therefore,

$$I(\boldsymbol{r}) = \sum_{j=1}^{N} I_j P_j(\boldsymbol{r}), \ \boldsymbol{r}\in l \tag{1.43}$$

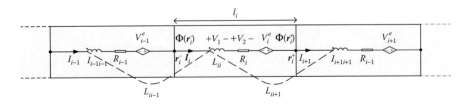

Figure 1.41 Current segments.

where $P_j(r) = \begin{cases} 1, & r \in l_j \\ 0, & r \notin l_j \end{cases}$ is the pulse function and I_j is the current flowing along the jth segment.

For both sides of Equation 1.39, by applying $\int_{l_i} \cdot dl$ along the ith segment (the vector dl has the same direction of I_i on ith current segment), we have:

$$j\omega\mu \sum_{j=1}^{N} \int_{l_i} \int_{l_j} G(r,r')dl' \cdot dl I_j + \left[\Phi(r_i^+) - \Phi(r_i^-)\right] + Z_s \frac{l_i}{w} I_i - \int_{l_i} E^e \cdot dl = 0 \quad (1.44)$$

where the vector dl' has the same direction of I_j on the jth current segment.

In the following, by using the Kirchhoff's current and voltage laws, an equivalent circuit can be derived from Equation 1.44.

1.2.5.2 Series Branch

Equation 1.44 can be rewritten as

$$j\omega \sum_{j=1}^{N} L_{ij} I_j + R_i I_i - V_i^e + \Phi(r_i^+) = \Phi(r_i^-), \text{ for } i = 1,\ldots,N \quad (1.45)$$

which results in:

$$V_1 + V_2 - V_i^e + \Phi(r_i^+) = \Phi(r_i^-) \quad (1.46)$$

where V_1, V_2, and V_i^e are shown in Figure 1.41, and

$$L_{ij} = \mu \iint_{l_i \ l_j} G(r,r') dl' \cdot dl, \quad R_i = Z_s \frac{l_i}{w} \quad \text{and} \quad V_i^e = \int_{l_i} E^e \cdot dl' \qquad (1.47)$$

$\Phi(r_i^+)$ and $\Phi(r_i^-)$ are the electric potentials at two ends of the ith current segment. Equation 1.46 follows the Kirchhoff's voltage law for the ith current segment, so the series equivalent circuit of the current segment can be obtained as in Figure 1.41.

1.2.5.3 Parallel Branch

It should be noted that due to the capacitive effect, there is current leakage at the node r_i^+ in Figure 1.41, so the currents flowing along the ith and $i + 1$th current segments are not equal. This leakage current can be described by using a parallel equivalent circuit. To do this, *M charge segments* are defined along the trace, as shown in Figure 1.42. Every charge segment is shifted from the current segment by about half length of the current segment, and usually, $M \neq N$. For the sake of clarity, in the following, subsection, i and j are used for current segments, whereas subscripts n and m are used for charge segments. The line electric charge density is expanded by using these charge segments as

$$q(r) = \sum_{n=1}^{M} q_n P_n(r), \ r \in l \qquad (1.48)$$

where $P_n(r) = \begin{cases} 1, & r \in n\text{th charge segment} \\ 0, & r \notin n\text{th charge segment} \end{cases}$ is the pulse function, and q_n is the charge density on the nth segment.

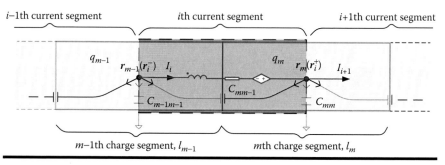

Figure 1.42 Charge segments.

In Figure 1.42, we chose the end of the current segment as the testing point of Equation 1.41, that is, $r_m = r_i^+$ and $r_{m-1} = r_i^-$. Substituting Equation 1.48 into Equation 1.41 and calculating its value at the testing point $r = r_m$, we can get:

$$\Phi_m = \Phi(r_m) = \sum_{n=1}^{M} p_{mn} Q_n, \text{ for } n = 1,\ldots,M \tag{1.49}$$

where p_{mn} is the potential coefficient.

$$p_{mn} = \frac{1}{l_n \varepsilon} \int_{l_n} G(r_m, r') dl' \tag{1.50}$$

and $Q_n = q_n l_n$ is the electric charge on nth charge segment. When l_n is small enough, for $m \neq n$,

$$p_{mn} = p_{nm} \approx \frac{1}{\varepsilon} G(r_m, r_n) \tag{1.51}$$

Equation 1.49 can be written in a matrix form as

$$\Phi = [P] \cdot Q \tag{1.52}$$

where $[P]$ is the potential matrix,

$$\Phi = \begin{bmatrix} \Phi_1 \\ \vdots \\ \Phi_M \end{bmatrix}, \text{ and } Q = \begin{bmatrix} Q_1 \\ \vdots \\ Q_M \end{bmatrix}.$$

The inverse of $[P]$ gives another $[B]$ matrix as

$$Q = [P]^{-1} \cdot \Phi = [B] \cdot \Phi \tag{1.53}$$

which results in:

$$Q_m = \sum_{n=1}^{M} B_{mn} \Phi_n, \text{ for } m = 1,\ldots,M \tag{1.54}$$

Φ_n is the electric potential of nth charge segment. It is also the voltage between the mth charge segment and the reference (usually the ground plane of the trace). $U_{mn} = \Phi_m - \Phi_n$ is the voltage between the mth and nth charge segments.

Equation 1.54 can be rewritten by replacing the potential Φ_n with the voltage U_{mn} and defining the capacitance C_{mn} between different charge segments as

$$Q_m = C_{mm}\Phi_m + \sum_{n=1,n\neq m}^{M} C_{mn}(\Phi_m - \Phi_n) = C_{mm}\Phi_m + \sum_{n=1,n\neq m}^{M} C_{mn}U_{mn}, \text{ for } m = 1,\ldots,M \quad (1.55)$$

where $C_{mm} = \sum_{n=1}^{M} B_{mn}$, and $C_{mn} = C_{nm} = -B_{mn} = -B_{nm}$ with $m \neq n$.

According to Equation 1.42, we have:

$$Q_m = q_m l_m = -\frac{1}{j\omega}\frac{\partial I(r')}{\partial l}\bigg|_{r'=r_m} \quad l_m \approx -\frac{1}{j\omega}\frac{I_{i+1} - I_i}{l_m}l_m = -\frac{1}{j\omega}(I_{i+1} - I_i) \quad (1.56)$$

where the mth charge segment is between the ith and $i+1$th current segments.
Comparing Equations 1.55 and 1.56, we get:

$$I_i = I_{i+1} + j\omega C_{mm}\Phi_m + \sum_{n=1,n\neq m}^{M} j\omega C_{mn}U_{mn} \quad (1.57)$$

Considering the Kirchhoff's current law at the mth charge segment and Equation 1.57, a parallel equivalent circuit of the charge segment can be obtained, as in Figure 1.42.

1.2.5.4 PEEC Circuit

Combining the equivalent circuits of Figures 1.41 and 1.42 together, the PEEC circuit of the trace shown in Figure 1.40 can be obtained, as in Figure 1.43. The RLC values in Figure 1.43 can be obtained from Equations 1.47 and 1.51. These integrals can be calculated by using the Gaussian quadrature formula. For $i = j$ and $m = n$, the Green's functions in the integrals of L_{ii} and p_{mm} are singular. When r or r_m is very close to r', the Green's functions are semi-singular. In these two cases,

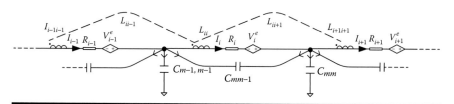

Figure 1.43   PEEC circuit.

the integrals of L_{ii} and Equation 1.50 are divided into a singular part, including $(1/(|r - r'|))$, and a non-singular part. The singular parts are analytically calculated by following the same procedure as in the appendix of Chapter 3, whereas the non-singular parts are numerically calculated by using the Gaussian quadrature formula [43].

This subsection demonstrates a one-dimensional PEEC algorithm. For most practical EMC problems, a 3D or 2D PEEC algorithm is required. The 3D or 2D PEEC algorithm also begins from the electric/magnetic field integral equation, where the line integral is replaced by the volume or surface integral. The meshing is very important for all PEEC algorithms. There are usually two kinds of mesh: the current mesh and the charge mesh, and their cells are shifted by about half cell length.

1.3 EMC Designs

In Section 1.1, EMC challenges for high-speed circuits are discussed. The EMC problem or failure is one of the major reasons of the product delay, which results in money loss. To solve these EMC problems, efficient and low-cost EMC control and design methods at the product design phase are required. After years of EMC research in both academic and industrial areas, there have been lots of EMC design rules for high-speed circuits. These include reducing the area formed by the current and its return current, avoiding parallel multitraces with a long distance, increasing the trace pitch to reduce coupling, using differential line instead of single-ended line for the high-speed signal, and so on. Owing to the ever-increasing clock frequency, all EMC control methods become frequency-dependent. For example, the shield box is used to prevent the electromagnetic field from the outside. However, at the resonant frequency of the shield box, it even couples stronger field from the outside. At the same time, there is no single solution for all kinds of EMC problems. We must consider the balance between different EMC targets, including SI, PI, and EMI. For the electromagnetic bandgap (EBG) design, the etching slot on PGPs can stop the noise propagation along them. However, this also introduces SI risk for the signal trace above the PGP and potential radiation from the slot.

Lots of EMC control methods had been published in literatures. Because of the complex and diverse nature of EMC problems, those methods are usually on case-by-case basis. In the following, some typical EMC control methods related to high-speed circuits are introduced, followed by the exploration about novel EMC structure designs.

1.3.1 Shield Box

Shield box/case is the most simple and effective solution for RE problems. It is widely employed in the system- and PCB-level EMC engineering. Figure 1.44a,b show the typical usage of shield boxes as the computer case and inside the smartphone,

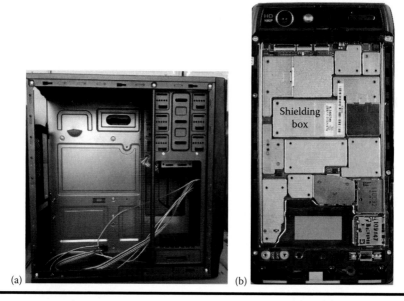

(a) (b)

Figure 1.44   **Shield boxes (a) as the computer case and (b) inside the smartphone.**

respectively. For a modern smartphone, various circuits are integrated inside a small space, especially the antennas for 2G to 4G, Bluetooth, WiFi, and so on, and are mounted closely to the noisy devices. The interference between them is so serious that the shield box is applied for almost every circuit model. The shield box is made of highly conductive metals and mounted to the surface of PCB by welding or by the conductive adhesive, and usually, shorting vias inside the PCB are connected to the shield box.

Owing to the slot/hole on the shield box for the heat-dissipation purpose and the air gap between the shield box and PCB, electromagnetic fields will leak into the shield box. Shielding effectiveness (SE) is used in practical EMC engineering to define shielding performance. At low frequency, the magnetic field has greater contribution to the EMI than the electric field, so the SE is defined as

$$SE_H(\text{dB}) = 20\log\frac{|H_1|}{|H_2|} \tag{1.58}$$

where $|H_1|$ and $|H_2|$ are the measured powers at the same location without and with the presence of the shield box, respectively. At a high frequency, both the magnetic and electric fields contribute to the EMI, so the SE is defined as

$$SE_P(\text{dB}) = 10\log\frac{P_1}{P_2} \tag{1.59}$$

where P_1 and P_2 are the measured power at the same location without and with the presence of the shield box, respectively.

Because of the resonant nature of the fields inside a shield box, the field inside the box shows a standing wave and is a function of the location where measurement is performed. Especially at the maximum value of the field standing wave, the calculated SE can be negative. To solve this problem, the Institute of Electrical and Electronics Engineers (IEEE's) testing standard [44] suggests sampling the field at various locations in the box and then taking the averaged field/power level to calculate SE or using a reverberation chamber with the mode-stirring or frequency-stirring to measure the SE.

1.3.2 Cross talk

Cross talk is a critical issue during the trace layout in a high-speed circuit. Based on the cross talk model in subsection 1.1.5, several methods are proposed to reduce the cross talk. Enlarging the trace-to-trace spacing can greatly reduce the coupling and then the cross talk between two traces. The 3W rule is widely used in the PCB trace layout, where the center-to-center distance between two adjacent traces should be equal to or larger than three times of the trace width and the distance between the PCB edge and the trace edge should be larger than three times of the trace width. However, the trace layout in modern PCB is denser than before, which makes it hard to fully follow the 3W rule.

To reduce the mutual inductance and capacitance between two traces, the length of parallel traces should be reduced. Sometimes, the jog layout shown in Figure 1.45a can be used to control the maximum parallel length. Other methods suggest the following: the traces in the adjacent layers should be routed at 90°, as shown in Figure 1.45b, which is helpful in reducing their inductive coupling; the usage of the low-permittivity dielectric substrate can also reduce the capacitive cross talk; a good impedance matching at the end of the trace will be helpful in reducing the magnitude of the cross talk, and so on. A guard line with ground vias can be placed between two signal traces to shield the coupling between them [45]. Based on this, a guard trace with a serpentine form was proposed to effectively reduce the FEXT voltage [46].

Equation 1.16 implies that FEXT of two traces is decided by the capacitive coupling factor $(C_m/C_s + C_m)$ and the inductive coupling factor (L_m/L_s). If the

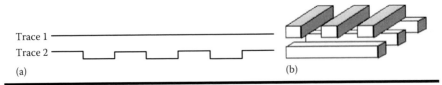

Figure 1.45 Methods to reduce the cross talk. (a) Jog layout and (b) routing traces by 90°.

traces locate in a homogeneous media, $(C_m/(C_s + C_m)) = (L_m/L_s)$, so that FEXT is zero. For the microstrip lines, the capacitive coupling factor is slightly smaller than the inductive coupling factor, because the dielectric constant of the surrounding air is smaller than that of the PCB dielectric material. This results in FEXT. Based on this analysis, some methods are proposed to composite the difference between the capacitive and inductive factors. The serpentine microstrip lines were proposed in [47] to increase the capacitive coupling factor, and an asymmetric stripline structure is designed with different dielectric constants to generate the stripline FEXT voltage that has a polarity opposite to that of the micro-stripline FEXT [48].

1.3.3 Differential Signaling

Differential signaling is a signal transmission technology. It is different from the single-ended signaling, in which the signal line together with a ground plane is employed; differential signaling uses two lines to transmit signals and responds to the electrical difference between the two signals rather than to the difference between a single line and the ground. In comparison with the single-ended trace, differential traces show several advantages, such as:

1. A differential circuit is insensitive to the noise resulting from the shared-ground or PGPs.
2. The currents flowing along two traces of one differential pair have the same amplitude and opposite phase. When the spacing between two traces is small enough, the radiated electric fields from each trace will also have the same magnitude and opposite polarization and then will cancel each other. This results in a low-EMI radiation.
3. For the noise coupled to or illuminated on two traces with the same magnitude and phase, they will also cancel each other at the input of the differential circuit. This gives a high electromagnetic immunity.

Owing to their advantages, differential pairs are widely used for the high-speed or high-frequency signal on PCBs, as shown in Figure 1.46, in twisted-pair and ribbon cables, and in connectors where the signals are very easy to radiate and also sensitive to noise. For example, for the 10-Gbit Ethernet circuit, four differential pairs running at 2.5 Gbit/s are employed.

For the loose coupling, where the center-to-center distance between two traces is larger than three times of the trace width, the characteristic impedance of the differential pair defined between two traces is equal to two times of the characteristic impedance of the single trace. The characteristic impedance of the differential pair decreases with the decreased pitch of two traces. The tight coupling between two traces of the differential pair is helpful in increasing its immunity and reducing its EMI.

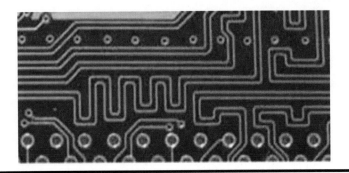

Figure 1.46 Differential signaling.

One major EMC concern about the differential pair is the common mode noise induced on it. Through two traces of the differential pair, the arriving of signals at different time (skew) will result in the distortion of the differential signal. The skew and distortion will convert part energy of the differential signal into the common mode noise, which has a stronger radiation. The skew results from the asynchronous signals, or the asymmetry of the differential pair in practical PCBs, so that the equivalent lengths of two traces are not equal.

The bend of the differential pair is a common asymmetric structure, as shown in Figure 1.47a, where the bend is equivalent to the two-conductor circuit. Since the inner trace of the bend is shorter than the outer trace, the field-traveling time along the inner trace is shorter than that of the outer trace. At the same time, the

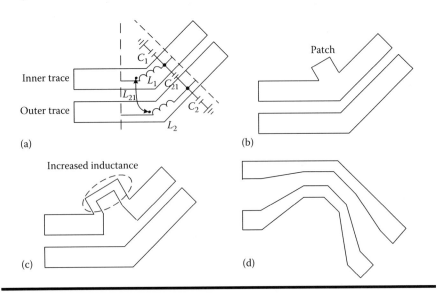

Figure 1.47 (a) Bend of the differential pair and its equivalent circuit, (b) compensation capacitance, (c) compensation inductance, and (d) tapered bend.

self-inductance L_1 and capacitance C_1 of the inner trace are smaller than the self-inductance L_2 and capacitance C_2 of the outer trace. Such asymmetry increases the differential-common mode conversion ratio. To solve this problem, some compensation structures are proposed. A compensation capacitance is implemented by a small patch attached to the inner trace at the bend to increase its self-capacitance, as shown in Figure 1.47b, and a compensation inductance is also proposed to increase the self-inductance of the inner trace, as shown in Figure 1.47c. Both of them can increase the field-traveling time along the inner trace [49]. At the same time, a tapered bend is also proposed in Figure 1.47d and optimized to reduce the differential-common mode conversion ratio at the bend [50]. For a general purpose, common mode filters are employed to eliminate the common mode along the differential pair, the details of which will be discussed in Chapter 5.

1.3.4 Via Stub

In a multilayered PCBs, the via stub is the residual part of the through-hole via after the via construction, as shown in Figure 1.48a,b. The via stub can be considered an open-ended transmission line connected in parallel to the signal trace. The stub impedance is frequency-dependent and will change the signal trace impedance at the stub connection. When the stub length is smaller than a quarter wavelength, the stub can be considered a capacitance. When the stub length is about a quarter wavelength (or at the resonant frequency of the stub), due to the impedance transformation of the transmission line, the stub can be considered a short circuit at the stub connection. This will result in large signal reflection, and the stub works like a stopband filter.

Figure 1.48a,b show a trace + via + via stub structure, where port 1 is defined at the end of the trace and port 2 is defined over the antipad. Port 2 is used to connect to the subminiature A (SMA) connector. Several shorting pins surrounding the via are used to connect all ground planes. They also contact with the outer conductor of the SMA when the SMA is mounted on the PCB surface. Figure 1.48c plots the $|S_{21}|$ of the trace. From this figure, we can see that at the resonant frequency of the stub, $|S_{21}|$ is greatly reduced.

To reduce the stub effect, several methods are used in the EMC design, including thinner boards (reducing the stub length and then increasing the resonant frequency of the stub) and backdrill (the stub will be drilled from the bottom of the board during the PCB fabrication). Another method is the blind or micro via. Unlike backdrill, the blind or micro via will not leave an air hole at the original position of the stub and has a better performance than the backdrill. However, both the backdrill and blind or micro via will increase the PCB cost. People also explore to use absorber material to absorb the power of the stub [51]. This makes the stub like a lossy or resistance-terminated transmission line, so that its reflection is reduced. The cost of this method is that it will introduce loss to the signal trace.

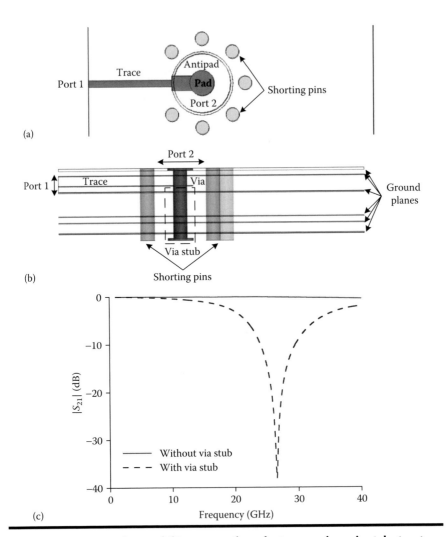

Figure 1.48 **(a) Top view and (b) cross section of a trace + via + via stub structure, and (c) the $|S_{21}|$ with and without the via stub.**

1.3.5 Silicon Loss

Silicon is used as a substrate of the interconnector in the TSV-based interposer. Because silicon has a larger conductivity than the substrate used in PCB, the IL and noise coupling of the interconnector in the interposer must be carefully considered, especially when the low-cost silicon with a low resistivity is employed. Some interconnector designs had been explored to reduce such IL. The basic idea of those designs is to reduce the strength of the electromagnetic fields penetrating into the substrate. Since the electromagnetic power consumed by the substrate is proportional to the square of the electric or magnetic field inside the substrate, those designs can reduce the signal loss and noise spreading inside the substrate. The guard ring, which serves as a ground line surrounding the device, is widely studied to reduce noise propagation; the coaxial TSV is proposed [52], where an additional cylindrical metal liner is placed around the original TSV, so that the electromagnetic leakage from the original TSV to the substrate is reduced.

Another way to reduce the current leakage from the horizontal interconnector or TSV to the substrate is to reduce the parasitic capacitance between the interconnector/TSV and the substrate. An air-gap liner between a copper TSV and the silicon substrate is proposed [53]. Since the permittivity of air is smaller than that of silicon, it can greatly reduce the effective TSV capacitance. The same idea is also used in horizontal interconnectors, where some part of the insulator layer between the interconnector and the substrate is etched.

In the following, the signal transmission coefficient of different interconnector configurations based on the same low-resistivity silicon is compared. Silicon substrate has the relative permittivity of 11.9 and the conductivity of 10 s/m, and SiO_2 is used as an insulator between copper traces and the substrate. Figure 1.49a shows the original GS traces. Figure 1.49b shows the same GS traces, except that there are metal grids (serve as the ground plane) inserted between two layers of SiO_2. In Figure 1.49c, the SiO_2 between the original G and S traces are etched. Figure 1.50 plots the $|S_{21}|$ of these three kinds of GS traces, and their electric field distributions at 10 GHz are shown in Figure 1.49. In Figure 1.50, the reference impedances used to calculate S parameters of three GS traces are equal to their characteristic impedances, so that the refection effect on S_{21} is removed. We can see that after adding the ground plane below GS, the electric field leakage into the substrate is greatly reduced, which results in the reduction of the IL. Etching the SiO_2 layer can also slightly reduce the IL.

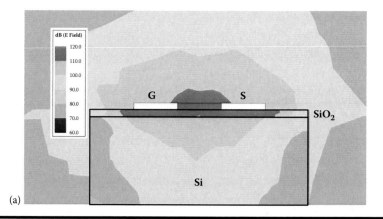

(a)

Figure 1.49 **The configuration and electric field distribution at 10 GHz for (a) original GS.** **(Continued)**

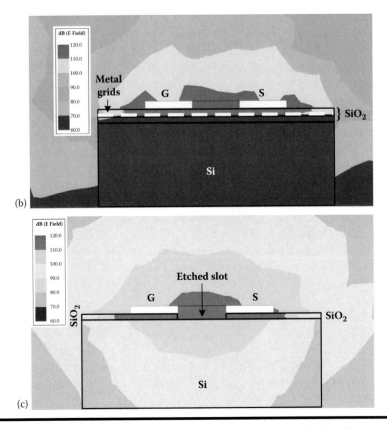

Figure 1.49 (Continued) The configuration and electric field distribution at 10 GHz for, (b) adding ground grids, and (c) etching the SiO_2 layer.

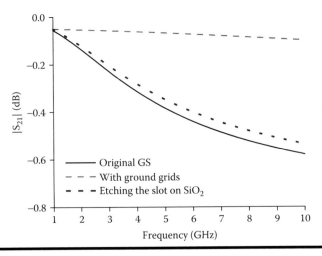

Figure 1.50 $|S_{21}|$ of three kinds of GS traces.

1.3.6 Electromagnetic Bandgap

Owing to the growing complex EMC problems of the high-speed circuit and ever-increasing EMC requirement, traditional EMC-control methods face a great challenge about the wide bandwidth, low cost, and compact size. There is a demand for the development of new structures and new materials for EMC engineering. As a new structure, the EBG technology offers promising alternatives to overcome the limitations of the traditional technology. Since its proposal until recently, EBG is one of the most rapidly advancing sectors in EMC research area.

Most of the EBG researches in EMC engineering focus on the PI design. Two typical EBG structures [54], the mushroom type and the coplanar type, are developed for this purpose, which are shown in Figure 1.51a,b respectively. They are periodic structures. Mushroom-type EBG is formed by embedding mushroom-shaped unit cells between the power and ground planes, whereas coplanar-type EBG is formed by etching slot on the power and/or ground planes and connecting every cell by a trace bridge. Their basic principle is that by using the patch, via, and slot, every unit cell can be equivalent to an *LC* circuit of a stopband filter. The whole EBG can be taken as a 2D array of stopband filters. They can eliminate the propagation of the parallel-plate modes and noise between the power and ground planes within their stopband. To quantitatively study the performance of the EBG, a 2D dispersion diagram, calculated by full-wave methods, is used, where all modes are solved for the unit cell with the periodic boundary conditions. From the dispersion diagram, the stopband and the slow-wave behavior of the EBG can be clearly observed.

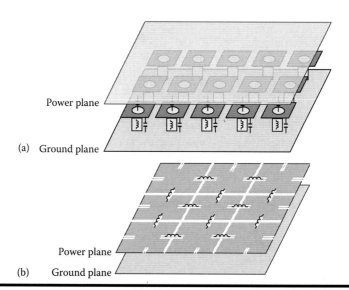

Figure 1.51 **(a) Mushroom-type and (b) coplanar-type EBG structures.**

Besides the PI design, EBG is used to eliminate the surface-wave propagation on the PCB [55]. Surface wave provides the noise coupling between different circuits mounted on the same PCB. It is also an awkward problem for the microstrip patch antenna array design, which can result in poor efficiency, large back lobe, and increase in antenna elements coupling. Surface waves can become dominant if a high dielectric constant and thick substrates are used. Since EBG can stop wave propagation between metal planes, based on the same idea, by etching slot on the metal layers of a PCB, it can be used to eliminate the wave propagation along its surface. This kind of EBG is also named as the high-impedance surface (HIS). It can be used to reduce the surface wave. Detailed analysis will be given in Chapter 7.

As a new technology, lots of researches efforts are made to improve its performance and real applications. A wider stopband of EBG is proposed to cover wide spectrum of the switching noise from digital circuits [56]. The miniaturization of EBG unit cell is required for the EBG to be used in a small space, such as the package. For mushroom-type EBG structures, double mushroom-shaped layers can increase the capacitance, as shown in Figure 1.51a, and reduce the unit cell size. For coplanar-type EBG structures, the bandwidth enhancement or the unit cell size reduction could be achieved by increasing the bridge inductance, as shown in Figure 1.51b. Recently, the tunable and multiband EBG designs have become hot research topics, since they provide a flexible filter, which finds application in radio frequency (RF) designs. The EBG technology represents a major breakthrough with respect to the traditional electromagnetic structure design, mainly due to its ability to guide and efficiently control electromagnetic waves. Many new structures will evolve from it.

1.3.7 Near-Field Scanning

With the increased clock frequency, the size of packages and ICs mounted on the PCB is comparable with the wavelength of interest. This makes them unwanted antennas on the PCB, and they become the major source of EMI. For example, measurements show that the application processor (AP) inside a smartphone generates strong interference to all kind of integrated antennas. To measure the EMI from packages or ICs, the TEM cell and anechoic chamber are employed. However, the TEM cell gives only an approximation of the total radiated power from the devices. It cannot fully describe the radiation properties of the EMI source, including the radiation pattern. The anechoic chamber is usually used to measure the far field of the device; however, the EMI between different circuits on a PCB is always within the near-field region.

Recently, there have been a great advancement of the near-field scanning technology, which makes it a powerful tool for analyzing the EMI problem of the high-speed circuit [57]. The basic idea of the near-field scanning technology is shown in

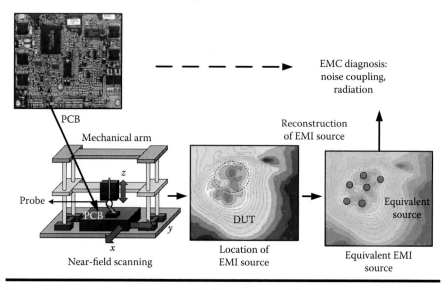

Figure 1.52   Basic idea of the near-field scanning technology.

Figure 1.52. Device under test (DUT) is equivalent to an equivalent EMI source (usually a dipole array). The equivalent source has the same radiation as the DUT but has a simpler structure. The radiated electromagnetic fields on a plane above the DUT are scanned by using the electric and/or magnetic probes, and the equivalent EMI source is obtained through the scanned fields by solving the related electromagnetic inverse problem. Finally, the equivalent EMI source is used for the EMC diagnosis, such as for estimating the noise coupling between different circuits on the same PCB and for analyzing the DUT radiation.

The major advantage of the near-field scanning technology over available EMI-measurement methods is that it provides a simple method to describe as much detailed radiation properties of the EMI source as possible, without knowing the details of the EMI source. For most cases, the structure of the EMI source, such as the AP inside the smartphone, is so complex that it is impossible to directly model its radiation in any commercial electromagnetic software. Moreover, due to the IP issue, usually, the EMC engineers cannot get the detailed physical structure of the EMI sources, such as the packages and ICs. Therefore, a simple and accurate equivalent EMI source is required. It is helpful in estimating the EMI risk of chips. If their noise coupling to other circuits is too large, then additional shield methods should be designed before they are mounted on the PCB.

The electric and magnetic infinitesimal dipole arrays are the mostly used equivalent EMI source due to their simplicity. In the free space, the radiated electromagnetic fields from an electric infinitesimal dipole located at the origin of a spherical coordinates are

$$
\begin{cases}
E_r = \eta \dfrac{I_0 l}{2\pi r^2} \cos\theta \left[1 + \dfrac{1}{jkr}\right] e^{-jkr} \\[2em]
E_\theta = j\eta \dfrac{kI_0 l}{4\pi r} \sin\theta \left[1 + \dfrac{1}{jkr} - \dfrac{1}{(kr)^2}\right] e^{-jkr} \\[2em]
H_\phi = j \dfrac{kI_0 l}{4\pi r} \sin\theta \left[1 + \dfrac{1}{jkr}\right] e^{-jkr} \\[2em]
E_\phi = H_r = H_\theta = 0
\end{cases}
\tag{1.60}
$$

where I_0 and l are the current and length of the electric dipole, respectively, and r is the distance between the observation point and the dipole. The radiated electromagnetic fields from a magnetic infinitesimal dipole located at the origin of a spherical coordinates are

$$
\begin{cases}
H_r = \dfrac{I_m l}{\eta 2\pi r^2} \cos\theta \left[1 + \dfrac{1}{jkr}\right] e^{-jkr} \\[2em]
H_\theta = j \dfrac{kI_m l}{\eta 4\pi r} \sin\theta \left[1 + \dfrac{1}{jkr} - \dfrac{1}{(kr)^2}\right] e^{-jkr} \\[2em]
E_\phi = -j \dfrac{kI_m l}{4\pi r} \sin\theta \left[1 + \dfrac{1}{jkr}\right] e^{-jkr} \\[2em]
H_\phi = E_r = E_\theta = 0
\end{cases}
\tag{1.61}
$$

where I_m and l are the magnetic current and length of the dipole, respectively.

When the scanned fields have both magnitude and phase information, Equations 1.60 and 1.61 result in a set of linear equations. $I_0 l$ and $I_m l$ of every dipole can be calculated by solving these equations, where dipoles are assumed to be evenly distributed on a plane. Since the scanned fields are obtained from the near-field region, they have large measurement error. The least-square and regularization methods are used to get an accurate result.

For the low-cost scanner, the scanned fields only have the magnitude information. In this case, from Equations 1.60 and 1.61, a set of non-linear equations is obtained. To solve these equations, optimization methods such as the genetic algorithm (GA) and differential evolution (DE) [58] are used. The locations of every dipole are determined by the GA or DE method.

It should be noted that Equations 1.60 and 1.61 are for free-space dipoles. They are used to approximately calculate the radiation from the DUT mounted on the PCB. The ground-plane effect can be considered by using the images of the dipoles. To accurately include the substrate effect of a multilayered PCB, the multilayered Green's functions [59] can be used to get the complex radiation formula, instead of using Equations 1.60 and 1.61.

The dipole array reconstruction of EMI source is demonstrated by using a microstrip line and a patch antenna, as shown in Figure 1.53. The microstrip line is excited at one end, and some of its radiated power is coupled at the input port of the patch antenna. The magnitudes of the radiated near-magnetic fields H_x and H_y from the microstrip line are scanned at the plane 5 mm above the microstrip line. An equivalent magnetic dipoles array is used to present the microstrip line, which is defined on a plane 1 mm above the microstrip line. The scanned magnetic fields are used to calculate the magnetic dipole array by using the DE method. After that, the dipoles together with the patch antenna are imported into the full-wave software to calculate the received power at the input port of the patch antenna, where the microstrip line is replaced by the equivalent dipoles.

Figure 1.54 plots the magnetic field magnitudes obtained from the direct full-wave simulation of the structure of Figure 1.53 (denoted by "full-wave simulation") and the equivalent magnetic dipoles array (denoted by "equivalent model"). The plane on which the magnetic fields are calculated is 5 mm above the microstrip line and the frequency is 1 GHz. Figure 1.55 shows the coupled power at the input port of the antenna. A good correlation can be observed.

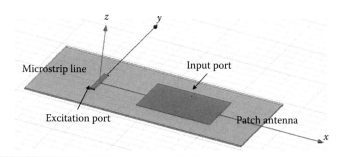

Figure 1.53 **A microstrip line and a patch antenna.**

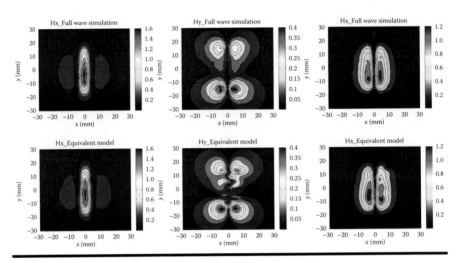

Figure 1.54 **Magnetic field magnitudes obtained from the direct full-wave simulation and the equivalent magnetic dipole array.**

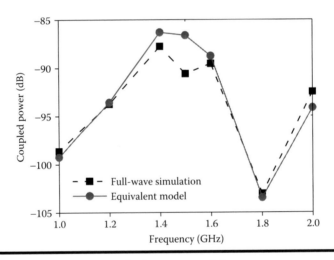

Figure 1.55 **Coupled power at the input port of the antenna.**

1.4 Organization of This Book

This book is organized into seven chapters. These chapters form a coherent unit. However, they are also structured in a way that every chapter is sufficiently self-contained to be read independently from the rest. The EMC-related acronyms are redefined in every chapter.

Chapter 1 introduces the major EMC issues related to high-speed PCBs and advanced packaging, their modeling methods, and the EMC design. After Chapter 1, the remainder of this book is divided into two parts: EMC modeling, which is explained in Chapters 2 to 4, and EMC design, which is elaborated in Chapters 5 to 7.

Chapter 2 presents the mode analysis method for PGPs and PGGs. This chapter begins with the derivation of the modal fields/eigenfunctions of PGPs from the 2D wave equation. After that, detailed examples are presented to explain the physical meaning of these modal fields, mode-control method, and equivalent circuit representation of the PGP impedance. Finally, we will introduce the solutions of modal fields by using the imaging method and the hybrid method and extend the modal field analysis method to PGGs, which are widely employed in advanced packaging.

Chapter 3 provides the integral equation solutions of the PDN, which can also be considered an extension of Chapter 1 to arbitrarily shaped PGPs. The integral equation methods are classified into two kinds according to their different Green's functions; we will discuss the 2D integral equation solution of the PDN, where the electromagnetic field is assumed uniform along the vertical direction and the Hankel function is used as the Green's function, and the 3D integral equation solution, where the electromagnetic field forms standing wave along the vertical direction and the dyadic Green's functions are employed. Finally, a combined equivalent networks analysis for PGPs with narrow slots is presented.

In Chapter 4, de-embedded and semi-analytical methods are proposed to model the vias used in PCBs and 3D/2.5D packages, respectively. For the through-hole vias in PCBs, we propose a non-equipotential transmission lines model to represent their parasitic circuits. For 3D/2.5D packages, a semi-analytical wideband modeling approach based on cylindrical mode expansion of electromagnetic fields is proposed to analyze the electrical properties of high-density TGV arrays.

Chapter 5 focuses on three common PCB-level EMC control methods: decoupling capacitor placement, common mode filter, and PCB-embedded structures/materials. We will analyze and optimize the Decap placement according to the modal field distribution of the PGPs proposed in Chapter 2, followed by the common mode filter design to suppress the common mode noise. Finally, the performance of PCB-embedded filters and absorbers is analyzed.

Chapter 6 explores the solutions to reduce the signal IL due to the lossy silicon substrate. For this purpose, a double-shielded interposer design with two metal layers directly contacting the silicon substrate surfaces for high-speed signal propagation along TSVs is analyzed. Next, two kinds of compact waveguide structures based on TGV technology are studied.

Chapter 7 discusses the applications of high-impedance surface structures and graphene films on EMC designs. The high-impedance surface is used to eliminate the surface-wave noise propagation inside a shield box, followed by the application of HIS on the performance enhancement of patch antennas working at WiFi frequency (2.45 GHz). Finally, the use of graphene as an absorbing film is discussed.

References

1. E. Sicard, J. F. Wu, R. J. Shen et al., Recent advances in electromagnetic compatibility of 3D-ICs-part I, *IEEE Electromagn. Compat. Mag.*, 4(4), 79–89, 2015.
2. *International technology roadmap for semiconductors*, 2016, http://www.itrs.net.
3. E. Bogatin, *Signal and Power Integrity-Simplified*, 2nd ed., Prentice Hall: Upper Saddle River, NJ, 2009.
4. M. Swaminathan and A. Engin, *Power Integrity Modeling and Design for Semiconductors and Systems.* Prentice Hall: Upper Saddle River, NJ, 2008.
5. X. C. Wei and E. P. Li, Integral-equation equivalent-circuit method for modeling of noise coupling in multilayered power distribution networks, *IEEE Trans. Microw. Theory Tech.*, 58(3), 559–565, 2010.
6. M. M. Waldrop, More than moore, *Nature*, 530, 145–147, 2016.
7. P. Garrou, Silicon interposers bridge to 3-D TSVs, *Semicond. Int.*, 2009.
8. P. Garrou, C. Bower, and P. Ramm, *Handbook of 3D Integration: Technology and Applications of 3D Integrated Circuits.* Wiley-VCH: Weinheim, Germany, 2008.
9. J. H. Lau, Evolution, challenge, and outlook of TSV (through-silicon via) and 3D IC/Si integration, in *International Symposium on Advanced Packaging Materials*, Xiamen, China, 2011, pp. 13–15.
10. J. Cong and Y. Zhang, Thermal via planning for 3-D ICs, in *International Conference on Computer-Aided Design*, San Jose, CA, 2005, pp. 744–751.
11. E. Wong and S. K. Lim, 3D floorplanning with thermal vias, in *Proceedings of Design Automation & Test in Europe Conference*, Munich, Germany, 2006, pp. 878–883.
12. J. Kim, E. Song, J. Cho, J. S. Pak, H. Lee, K. Park, and J. Kim, Through silicon via equalizer, in *18th IEEE Conference on Electrical Performance of Electronic Packaging*, Portland, OR, 2009, pp. 13–16.
13. C. Kim and Y. K. Yoon, High frequency characterization and analytical modeling of through glass via (TGV) for 3D thin-film interposer and MEMS packaging, in *IEEE 63rd Electronic Components and Technology Conference*, Las Vegas, NV, 2013, pp. 1385–1391.
14. V. Sridharan, M. Swaminathan, and T. Bandyopadhya, Enhancing signal and power integrity using silicon interposer, *IEEE Micro. Comp. Lett.*, 21(11), 598–600, 2011.
15. K. Koo, S. Lee, and J. Kim, Vertical noise coupling on wideband low noise amplifier from on-chip switching-mode DC-DC converter in 3D-IC, in *8th International Workshop on Electromagnetic Compatibility of Integrated Circuits*, Dubrovnik, Croatia, 2011, pp. 35–40.
16. J. H. Cho, J. J. Shim, E. H. Song, J. S. Pak, J. H. Lee, H. D. Lee, K. W. Park, and J. Kim, Active circuit to through silicon via (TSV) noise coupling, in *18th IEEE Conference on Electrical Performance of Electronic Packaging and Systems*, Portland, OR, 2009, pp. 97–100.
17. S. B. Dhia, M. Ramdani, and E. Sicard, *Electromagnetic Compatibility of Integrated Circuits—Techniques for Low Emission and Susceptibility.* Springer: Berlin, Germany, 2006.
18. H. B. Bakoglu, *Circuits, Interconnections, and Packaging for VLSI.* Addison Wesley: Upper Saddle River, NJ, 1990.
19. X. C. Wei, E. P. Li, E. X. Liu, and R. Vahldieck, Efficient simulation of power distribution network by using integral equation and modal decoupling technology, *IEEE Trans. Microw. Theory Tech.*, 56(10), 2277–2285, 2008.

20. Y. Brian, *Digital Signal Integrity-modelling and Simulation with Interconnectors and Packages*. Prentice Hall: Upper Saddle River, NJ, 2001.
21. E. P. Li, X. C. Wei, A. C. Cangellaris, E. X. Liu, Y. J. Zhang, M. D'Amore, J. Kim, and T. Sudo, Progress review of electromagnetic compatibility analysis technologies for packages, printed circuit boards, and novel interconnects, *IEEE Trans. Electromagn. Compat.*, 52(2), 248–265, 2010.
22. J. E. Bracken, S. Polstyanko, I. Bardi, A. Mathis, and Z. J. Cendes, Analysis of system-level electromagnetic interference from electronic packages and boards, in *Proceedings of 14th Conference on Electrical Performance of Electronic Packaging*, Austin, TX, 2005, pp. 183–186.
23. W. K. Gwarek, Analysis of an arbitrarily-shaped planar circuit a time-domain approach, *IEEE Trans. Microw. Theory Tech.*, 33(10), 1067–1072, 1985.
24. A. E. Engin, W. John, G. Sommer, W. Mathis, and H. Reichl, Modeling of striplines between a power and a ground plane, *IEEE Trans. Adv. Packag.*, 29(3), 415–426, 2006.
25. L. Tsang, H. Chen, C. C. Huang, and V. Jandhyala, Modeling of multiple scattering among vias in planar waveguides using Foldy-Lax equations, *Micro. Opt. Technol. Lett.*, 31(3), 201–208, 2001.
26. Z. Z. Oo, E. P. Li, X. C. Wei, E. X. Liu, Y. J. Zhang, and L. W. Li, Hybridization of the scattering matrix method and modal decomposition for analysis of signal traces in a power distribution network, *IEEE Trans. Electromagn. Compat.*, 51,(3), 784–791, 2009.
27. Z. Z. Oo, E. X. Liu, E. P. Li, X. C. Wei, Y. Zhang, M. Tan, L. W. Li, and R. Vahldieck, A semi-analytical approach for system-level electrical modeling of electronic packages with large number of vias, *IEEE Trans. Adv. Packag.*, 31(2), 267–274, 2008.
28. E. X. Liu, E. P. Li, Z. Z. Oo, X. C. Wei, Y. J. Zhang, and R. Vahldieck, Novel methods for modeling of multiple vias in multilayered parallel-plate structures, *IEEE Trans. Microw. Theory Tech.*, 57(7), 1724–1733, 2009.
29. C. J. Ong, D. Miller, L. Tsang, B. Wu, and C. C. Huang, Application of the Foldy-Lax multiple scattering method to the analysis of vias in ball grid arrays and interior layers of printed circuit boards, *Micro. Opt. Technol. Lett.*, 49(1), 225–231, 2007.
30. K. J. Han and M. Swaminathan, Polarization mode basis functions for modeling insulator-coated through-silicon via (TSV) interconnections, in *IEEE Workshop on Signal Propagation on Interconnects*, Strasbourg, France, 2009, pp. 1–4.
31. G. Katti, M. Stucchi, K. D. Meyer, and W. Dehaene, Electrical modeling and characterization of through silicon via for three-dimensional ICs, *IEEE Trans. Electron Dev.*, 57(1), 256–262, 2010.
32. E. X. Liu, H. M. Lee, X. C. Wei, and E. P. Li, Different designs of TSVs for 3D IC: signal integrity analysis with cascaded scattering matrix, in *IEEE Electrical Design of Advanced Packaging Systems Symposium*, Hangzhou, China, 2011, pp. 1–4.
33. J. Cho, E. Song, K. Yoon et al., Modeling and analysis of through silicon via (TSV) noise coupling and suppression using a guard ring, *IEEE Trans. Comp. Packag. Manufact. Technol.*, 1(2), 220–233, 2011.
34. K. H. Yoon, G. W. Kim, W. J. Lee, T. G. Song, and J. H. Kim, Modeling and analysis of near-end crosstalk between TSVs, metal interconnects and RDLs in TSV-based 3D IC and silicon interposer, in *11th IEEE Electronics Packaging Technology Conference*, Singapore, 2009.

35. C. Xu, R. Suaya, and K. Banerjee, Compact modeling and analysis of through-si-via-induced electrical noise coupling in three-dimensional ICs, *IEEE Trans. Electron Dev.*, 58(11), 4024–4034, 2011.
36. G. X. Luo, E. P. Li, X. C. Wei, X. Cui, and R. Hao, PDN impedance modeling for multiple through vias array in doped silicon, *IEEE Trans. Electromagn. Compat.*, 56(5), 1202–1209, 2014.
37. F. E. Terman, *Radio Engineer's Handbook*. McGraw-Hill Companies: New York, 1943.
38. A. E. Ruehli, Equivalent circuit models for three-dimensional multiconductor systems, *IEEE Trans. Microw. Theory Tech.*, 22(3), 216–221, 1974.
39. G. Antonini, A. Orlandi, A. E. Ruehli, J. Ekman, and J. Delsing, *PEEC Development Road Map*, 2016 http://pure.ltu.se/.
40. *Moss Bay EDA*, 2016 http://www.mossbayeda.com.
41. K. Nabors and J. White, Multipole-accelerated 3-D capacitance extraction algorithm for structures with conformal dielectrics, in *29th ACM/IEEE Design Automation Conference*, Los Alamitos, CA, 1999, pp. 710–715.
42. M. Kamon, M. J. Tsuk, and J. K. White, FASTHENRY: A multipole-accelerated 3-D inductance extraction program, *IEEE Trans. Microw. Theory Technol.*, 42(9), 1750–1758, 1994.
43. R. D. Graglia, On the numerical integration of the linear shape functions times the 3-D green's function or its gradient on a plane triangle, *IEEE Trans. Antennas Propagat.*, 41(10), 1448–1455, 1993.
44. *IEEE Standard Method for Measuring the Shielding Effectiveness of Enclosures and Boxes Having all Dimensions between 0.1 m and 2 m*, IEEE Std 299.1-2013, 2013.
45. L. Zhi, W. Qiang, and S. Changsheng, Application of guard traces with vias in the RF PCB layout, in *3rd International Symposium on Electromagnetic Compatibility*, Beijing, China, 2002, pp. 771–774.
46. K. Lee, H. B. Lee, H. K. Jung, J. Y. Sim, and H. J. Park, A serpentine guard trace to reduce the far-end crosstalk voltage and the crosstalk induced timing jitter of parallel microstrip lines, *IEEE Trans. Adv. Packag.*, 31(4), 809–817, 2008.
47. K. Lee, H. K. Jung, H. J. Chi, H. J. Kwon, J. Y. Sim, and H. J. Park, Serpentine microstrip lines with zero far-end crosstalk for parallel high-speed DRAM interfaces, *IEEE Trans. Adv. Packag.*, 33(2), 552–558, 2010.
48. M. H. Cheng, Y. J. Tung, D. Y. T. Lai, and C. N. Pai, Microstrip lines far-end crosstalk cancellation using striplines in hybrid PCB structure, in *7th Asia-Pacific International Symposium on Electromagnetic Compatibility*, Shenzhen, China, 2016, pp. 576–579.
49. C. P. Huang, D. B. Lin, Y. C. Chen, H. N. Ke, and W. S. Liu, Signal integrity improvements of bended coupled lines by using miniaturized capacitance and inductance compensation structures, in *7th Asia-Pacific International Symposium on Electromagnetic Compatibility*, Shenzhen, China, 2016, pp. 22–24.
50. C. Y. Wang, K. Iokibe, and Y. Toyota, Design methodology of tightly coupled asymmetrically tapered bend for high-density mounting in differential transmission lines, in *7th Asia-Pacific International Symposium on Electromagnetic Compatibility*, Shenzhen, China, 2016, pp. 463–465.
51. S. W. Huang, K. Xiao, and B. Lee, Electromagnetic wave absorption technology for stub effects mitigation, in *DesignCon*, Santa Clara, CA, 2016.

52. Z. Xu and J. Q. Lu, Three-dimensional coaxial through-silicon-via (TSV) design, *IEEE Electron Device Lett.*, 33(10), 1441–1443, 2012.
53. H. Oh, P. A. Thadesar, G. S. May, and M. S. Bakir, Low-loss air-isolated through-silicon vias for silicon interposers, *IEEE Electron Device Lett.*, 26(3), 168–170, 2016.
54. T. L. Wu, H. H. Chuang, and T. K. Wang, Overview of power integrity solutions on package and PCB: decoupling and EBG isolation, *IEEE Trans. Electromagn. Compat.*, 52(2), 346–356, 2010.
55. D. Sievenpiper, L. Zhang, R. F. J. Broas, N. G. Alexopolous, and E. Yablonovitch, High-impedance electromagnetic surfaces with a forbidden frequency band, *IEEE Trans. Microw. Theory Tech.*, 47(11), 2059–2074, 1999.
56. T. K. Wang, C. Y. Hsieh, H. H. Chuang, and T. L. Wu, Design and modeling of a stopband-enhanced EBG structure using ground surface perturbation lattice for power/ground noise suppression, *IEEE Trans. Microw. Theory Tech.*, 57(8), 2047–2054, 2009.
57. J. Pan, X. Gao, and J. Fan, Far-field prediction by only magnetic near fields on a simplified Huygens's surface, *IEEE Trans. Electromagn. Compat.*, 57(4), 693–701, 2015.
58. J. Li, X. C. Wei, and J. Li, Near-field measurements based source reconstruction approach for radiated emissions prediction, in *IEEE International Symposium on Electromagnetic Compatibility*, Ottawa, Canada, July 25–29, 2016.
59. M. I. Aksun, A robust approach for the derivation of closed-form Green's function, *IEEE Trans. Microw. Theory Tech.*, 44(5), 651–658, 1996.

Chapter 2

Modal Field of Power–Ground Planes and Grids

Power distribution network (PDN) is employed to supply power for circuits. It is an important and essential part for circuit integration at all levels: printed circuit boards (PCBs), packages, and even chips. To reduce the high-frequency impedance of the PDN and provide shielding between different substrate layers, power–ground planes (PGPs) or power–ground grids (PGGs) are widely used as a typical structure of PDN in multilayer PCBs, packages, and chips [1,2].

The rapid growth and convergence of digital computing and wireless communication have been driving the semiconductor industry to integrate more and more circuits on PCBs and into one single package. At the same time, the voltage supply level is continuously reduced with the ever-increasing working frequency. These make the electromagnetic compatibility (EMC), including the signal integrity and power integrity, a very critical issue for the successful design of PGPs and PGGs [3]. Considering the harmonics of the increased clock frequency, the noise spectrum on PCBs and inside the package will cover very high frequency. With this ever-decreasing wavelength, the electromagnetic wave fluctuation inside PGPs and PGGs cannot be ignored. This requires us to use the electromagnetic and microwave theory in all aspects of future circuit system, including modeling, design, and testing. We need to study the modal behaviors of the electric and magnetic fields, instead of the voltage and current for high-speed, high-power, and high-density electric circuit.

The demand for an accurate and efficient simulation of such complex EMC problems strongly motivates the development of more advanced electromagnetic modeling algorithms. The full-wave three-dimensional (3D) methods, such as finite element method (FEM) and finite-difference time-domain (FDTD) method, can give an accurate simulation result. However, their application is limited to small and simple geometry, due to their computational cost. Although the overall size of PGPs and PGGs is small enough to apply those full-wave methods, the high aspect ratio and the dense and small vias result in a huge mesh. This makes these full-wave methods very expensive in terms of the computing time. To avoid the computational cost of these full-wave methods, more efficient two-dimensional (2D) methods are proposed to simulate the signal and power integrity problem. A multilayered finite-difference method was proposed in [4]. The 2D FEM is also used to simulate PGPs [5]. An analytical-numerical method is proposed in [6].

In this chapter, we give a review of the state of the art of the mode analysis method for PGPs and PGGs. This method is easy to implement and usually needs fewer computing resources than the full-wave methods. Most important, the mode analysis can give an intuitive physical understanding of the electromagnetic behavior of PGPs and PGGs, including their resonances, equivalent circuits, and impedance control. It is much helpful for practical EMC engineers, since it provides "how" knowledge, which the full-wave method cannot provide.

This chapter will begin with the derivation of the modal fields/eigenfunctions of PGPs from a 2D wave equation, where different boundary conditions are considered (open, short, and hybrid). Analytical formula expanded by using eigenfunctions is obtained for PGPs with the rectangular shape. After that, detailed examples are presented to explain the physical meaning of these modal fields, mode control method, and equivalent circuit representation of the PGP impedance in the practical PGPs design. Finally, we will introduce the solution of modal field by using the imaging method and the hybrid method [7,8] and extend the modal field analysis method to PGGs; this method is widely employed in advanced packaging and serves as PGPs in PCBs [9].

It should be noted that the mode analysis method discussed in this chapter can also be extended to applications of microstrip/strip patch-based radio frequency (RF) structure and component design. Recently, there has been an increasing interest in the design of microstrip/strip patch-based antenna, filter, metamaterials, frequency selective surface (FSS), electromagnetic bandgap (EBG) [10], and so on. Most of them work on certain resonant frequency, so the mode analysis is the fundamental method to better understand and design those RF structures and components. The materials in this chapter are helpful for the novel RF structures and component designs.

2.1 Wave Equation and Its Solution by Using the Green's Function

2.1.1 Two-Dimensional Wave Equation

A typical PGP pair with an arbitrary shape is shown in Figure 2.1, where the metal planes can be power or ground planes employed in PCBs and packages and a substrate with the thickness of d is sandwiched between the metal planes. There can be power–ground traces connected to and signal traces passing through the PGPs. Only the electric field E and magnetic field H inside the substrate are considered here; they satisfy the following Maxwell's curl equation:

$$\nabla \times H = j\omega\varepsilon E + J_{Sub} + J_{Ext} \tag{2.1}$$

where ε is the electric permittivity of the substrate and ω is the angular frequency. J_{Ext} is the excitation current density, and it flows through the area Δs of the antipad. J_{Ext} comes from the return current of the signal/power–ground trace going through the antipad in Figure 2.1. J_{Sub} is the current density inside the substrate, which is related to E as

$$J_{Sub} = \sigma E \tag{2.2}$$

where σ is the electrical conductivity of the substrate.

Substituting Equation 2.2 into 2.1, we get:

$$\nabla \times H = j\omega\tilde{\varepsilon}E + J_{Ext} \tag{2.3}$$

where $\tilde{\varepsilon}$ is the *complex* permittivity of the substrate, with $\tilde{\varepsilon} = \varepsilon(1 - j\tan\delta)$, and δ is the loss angle, with $\tan\delta = (\sigma/\omega\varepsilon)$. In this way, the substrate loss is included in $\tilde{\varepsilon}$. If considering the metal plane loss, $\tilde{\varepsilon}$ can be further rewritten as $\tilde{\varepsilon} = \varepsilon[1 - j(\tan\delta + t/d)]$ with t being the skin depth of the metal planes. Since,

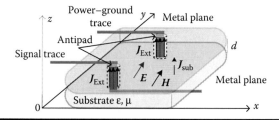

Figure 2.1   Typical structure of PGPs.

usually, copper with a high conductivity is employed for PGPs, the metal loss is neglected in the following:

Another Maxwell's curl equation is:

$$\nabla \times \boldsymbol{E} = -j\omega\mu\boldsymbol{H} \tag{2.4}$$

where μ is the magnetic permeability of the substrate.

Applying $\nabla \times$ on both sides of Equation 2.4 and considering Equation 2.3, we can get:

$$\nabla \times \nabla \times \boldsymbol{E} = k^2 \boldsymbol{E} - j\omega\mu\boldsymbol{J}_{\text{Ext}} \tag{2.5}$$

where the complex wave number is

$$k = \omega\sqrt{\tilde{\varepsilon}\mu} \tag{2.6}$$

Assuming that there are no free charges inside the substrate, $\nabla \cdot \boldsymbol{E} = 0$. We get:

$$\nabla \times \nabla \times \boldsymbol{E} = \nabla\nabla \cdot \boldsymbol{E} - \nabla^2 \boldsymbol{E} = -\nabla^2 \boldsymbol{E} \tag{2.7}$$

Therefore, Equation 2.5 is rewritten as

$$(\nabla^2 + k^2)\boldsymbol{E} = j\omega\mu\boldsymbol{J}_{\text{Ext}} \tag{2.8}$$

Above equation is a 3D wave equation, the solution of which is much complex, considering both $\boldsymbol{E}$ and $\boldsymbol{J}_{\text{Ext}}$ have three components. However, according to the following approximations, we can reduce Equation 2.8 to a 2D wave equation.

1. The thickness of the substrate d is very smaller compared with the working wavelength, so that the fields inside the substrate (between the two metal planes, as shown in Figure 2.1) can be assumed to be uniform along the z direction, and hence, we have $\partial/\partial z = 0$ for the electric and magnetic fields.
2. For most PCBs and packages, the PGPs are made of copper with a high conductivity. They can be taken as perfect electric conductor (PEC), so we have:

$$E_x = E_y = H_z = 0 \tag{2.9}$$

on the surface of metal planes. Since the field inside the substrate is uniform along the z direction, Equation 2.9 holds for the whole substrate.

In summary, inside the substrate, we have:

$$\boldsymbol{E}(\boldsymbol{r}) = \hat{z}E_z(x, y) \tag{2.10}$$

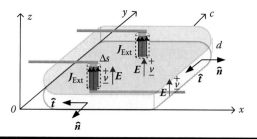

Figure 2.2 Voltage and current definitions of PGPs.

$$H(r) = \hat{x}H_x(x, y) + \hat{y}H_y(x, y) \tag{2.11}$$

$$J_{\text{Ext}}(r) = \hat{z}J_{\text{Ext},z}(x, y),\ r = x\hat{x} + y\hat{y} \tag{2.12}$$

Therefore, Equation 2.8 is reduced to the following 2D wave equation:

$$(\nabla_t^2 + k^2)E_z(x, y) = j\omega\mu J_{\text{Ext},z}(x, y) \tag{2.13}$$

with $\nabla_t^2 = (\partial^2/\partial x^2) + (\partial^2/\partial y^2)$.

Equation 2.13 is in the form of electric fields; however, to connect this PGPs model to other circuit model in PCBs and packages, a wave equation in the form of voltage and current is more preferred. As shown in Figure 2.2, the voltage at any location between two metal planes is defined as

$$v = -E_z d \tag{2.14}$$

Substituting Equation 2.14 into Equation 2.13, we get the wave equation in the form of current and voltage:

$$(\nabla_t^2 + k^2)v(r) = -j\omega\mu df(r) \tag{2.15}$$

where $f(r) = I_{\text{Ext}}(r)/\Delta s$ and I_{Ext} is the total current flowing through the area of the antipad. I_{Ext} is the return current of the signal current flowing through the antipad area.

2.1.2 Boundary Conditions

In order to solve the wave Equation 2.15, suitable boundary condition must be applied. Considering the scenario in practical PCBs and packages design, two boundary conditions are discussed in the following subsections.

2.1.2.1 Open Boundary

Open boundary means that the PCB is placed in free space, and there is no shorting vias along its contour. The open boundary can be taken as a perfect magnetic conductor (PMC) from the view of the electromagnetic field. For the PMC boundary, we have: $\hat{n}(r) \cdot E(r)\big|_{r \in C} = \hat{n}(r) \times H(r)\big|_{r \in C} = 0$, where C is the contour along PCBs and $\hat{n}$ is the outer unit vector vertical to the contour, as shown in Figure 2.2.

$\hat{n} \cdot E = 0$ can be easily achieved by considering that the electric field only has z component.

According to Equation 2.4, we have:

$$
\begin{aligned}
\hat{n} \times H &= \frac{1}{-j\omega\mu} \hat{n} \times \nabla \times E = \frac{1}{-j\omega\mu} \hat{n} \times \nabla \times (\hat{z}E_z) \\
&= \frac{1}{-j\omega\mu} \hat{n} \times [\nabla E_z \times \hat{z} + E_z \nabla \times \hat{z}] \\
&= \frac{1}{-j\omega\mu} \hat{n} \times (\nabla E_z \times \hat{z}) \\
&= \frac{1}{-j\omega\mu} [\nabla E_z(\hat{z} \cdot \hat{n}) - \hat{z}(\nabla E_z \cdot \hat{n})] \\
&= \frac{\hat{z}}{j\omega\mu} \nabla E_z \cdot \hat{n} \\
&= \frac{\hat{z}}{j\omega\mu} \cdot \frac{\partial E_z}{\partial n}
\end{aligned}
\tag{2.16}
$$

Since $\hat{n}(r) \times H(r)\big|_{r \in C} = 0$, we have $\partial E_z(r)/\partial n\big|_{r \in C} = 0$, which is an open-boundary condition for Equation 2.13. Hence, $\partial v(r)/\partial n\big|_{r \in C} = 0$ is an open-boundary condition for Equation 2.15.

2.1.2.2 Short Boundary

In most practical PCBs designs, lots of shorting vias are placed along the contour of PGPs to eliminate the unwanted radiation from them. For this case, the contour can be taken as a short boundary. The short boundary can be taken as a PEC from the view of the electromagnetic field. When the contour of PCB is a PEC boundary, we have: $\hat{n}(r) \cdot H(r)\big|_{r \in C} = \hat{n}(r) \times E(r)\big|_{r \in C} = 0$. Since $\hat{n} \times E = \hat{n} \times (\hat{z}E_z) = \hat{t}E_z$, where $\hat{t} = \hat{n} \times \hat{z}$ is the unit vector tangential to the contour, so $E_z(r)\big|_{r \in C} = 0$ becomes a short-boundary condition for Equation 2.13. Hence, $v(r)|_{r \in C} = 0$ is a short-boundary condition for Equation 2.15.

Besides the open and short boundaries, there can be impedance boundary for the PGPs. In fact, due to the radiation from the contours of the PGP, the PGP can be considered terminated with the frequency-dependent radiation impedance. One solution of the wave equation under this impedance boundary is that, first, get the equivalent network (will be discussed in the following subsections) of the PGPs, with the ports defined along its contour, and then, the radiation impedance can be calculated from the radiation field from a path antenna; finally, this radiation impedance is connected to the equivalent network of the PGPs and other circuits/devices to perform system simulation. For practical EMC analysis, an easy modeling method is to consider the radiation into the loss of the substrate, and the analytical formula can be obtained [11].

2.1.3 Solution of Wave Equation

Earlier, we obtained the wave Equation 2.15 and its boundary conditions. The voltage v in the wave equation can be solved by using the Green's function under the same boundary conditions. This solution procedure includes two steps: first, to express the voltage v by using the integral of the Green's functions, and then, to express the Green's function by using the summation of eigenfunctions. These two steps are introduced in the following subsections.

2.1.3.1 Green's Function

Here, we rewrite the wave equation and its open boundary condition as

$$(\nabla^2 + k^2)v(r) = -j\omega\mu df(r) \tag{2.17}$$

$$\left.\frac{\partial v(r)}{\partial n}\right|_{r \in C} = 0 \tag{2.18}$$

The corresponding Green's function and its boundary condition are defined as

$$(\nabla^2 + k^2)G(r,r') = -j\omega\mu d\delta(r - r') \tag{2.19}$$

$$\left.\frac{\partial G(r,r')}{\partial n}\right|_{r \in C} = 0 \tag{2.20}$$

where $\delta(r - r')$ is the Dirac delta function. $r = x\hat{x} + y\hat{y}$ is the observation point and $r' = x'\hat{x} + y'\hat{y}$ represents the source point.

v can be expressed by using the Green's function as

$$v(r) = \int_S G(r,r') f(r') ds' \tag{2.21}$$

where S is the area of the metal plane, as shown in Figure 2.2.

Proof:

According to the Green's second identity and surface-divergence theorem, we have:

$$\int_S \left(\phi \nabla_t^2 \psi - \psi \nabla_t^2 \phi \right) ds = \int_C \left(\phi \nabla_t \psi - \psi \nabla_t \phi \right) \cdot \hat{n} dl \tag{2.22}$$

where C is the contour of the area S, as shown in Figure 2.2, $\hat{n}$ is the outer unit vector normal to the contour C.

Let $\phi = G$ and $\Psi = v$, we get:

$$\int_S \left(G \nabla_t^2 v - v \nabla_t^2 G \right) ds = \int_C \left(G \nabla_t v - v \nabla_t G \right) \cdot \hat{n} dl \tag{2.23}$$

By replacing $\nabla_t^2 v$ and $\nabla_t^2 G$ with Equations 2.17 and 2.19, respectively, and considering the boundary condition $\nabla_t v(r) \cdot \hat{n}\big|_{r \in C} = \partial v(r)/\partial n\big|_{r \in C} = 0$ and $\nabla_t G(r,r') \cdot \hat{n}\big|_{r \in C} = \partial G(r,r')/\partial n\big|_{r \in C} = 0$, we have

$$\int_S \left\{ G(r,r')[-j\omega\mu df(r) - k^2 v(r)] - v(r)[-j\omega\mu d\delta(r-r') - k^2 G(r,r')] \right\} ds = 0 \tag{2.24}$$

$$\int_S -j\omega\mu df(r)G(r,r')ds + j\omega\mu dv(r') = 0 \tag{2.25}$$

Since the Green's function is symmetric [i.e., $G(r,r') = G(r,r')$], we get:

$$v(r) = \int_S G(r,r') f(r') ds' \tag{2.26}$$

2.1.3.2 Eigenfunctions

The above Green's function can be calculated by the summation of eigenfunctions. The eigenfunctions $\phi_n(r)$ can be defined as follows:

$$(\nabla^2 + k_n^2)\phi_n(\mathbf{r}) = 0 \tag{2.27}$$

where k_n^2 are the corresponding eigenvalues and n denotes all needed indices defining a certain $\phi_n(\mathbf{r})$. The non-trivial solution of Equation 2.27 is present only for an infinite number of discrete values of k_n^2. The eigenfunction is also the modal voltage/field distribution for the problem of PGPs, which we will discuss in details later.

The boundary condition of the eigenfunctions is same as that of the Green's function:

$$\left.\frac{\partial \phi_n(\mathbf{r})}{\partial n}\right|_{r \in C} = 0 \tag{2.28}$$

Given that the eigenfunctions are orthonormal, they must satisfy:

$$\int_S \phi_n^*(\mathbf{r})\phi_m(\mathbf{r})ds = \begin{cases} 1, & \text{if } n = m \\ 0, & \text{otherwise} \end{cases} \tag{2.29}$$

where $*$ symbolizes a complex conjugate.

Using $\phi_n(\mathbf{r})$ as the basic functions, the Green's function defined in Equations 2.19 and 2.20 can be expanded as

$$G(\mathbf{r},\mathbf{r}') = \sum_n A_n(\mathbf{r}')\phi_n(\mathbf{r}) \tag{2.30}$$

In the following equation, the unknown coefficient A_n is determined.

Substituting Equation 2.30 into Equation 2.19, we get:

$$\sum_n A_n(\mathbf{r}')\left[\nabla^2\phi_n(\mathbf{r}) + k^2\phi_n(\mathbf{r})\right] = -j\omega\mu d\delta(\mathbf{r} - \mathbf{r}') \tag{2.31}$$

Considering Equation 2.27,

$$\sum_n A_n(\mathbf{r}')(k^2 - k_n^2)\phi_n(\mathbf{r}) = -j\omega\mu d\delta(\mathbf{r} - \mathbf{r}') \tag{2.32}$$

Integrating both sides with $\phi_n^*(\mathbf{r})$ and considering the orthogonality of $\phi_n(\mathbf{r})$, we get

$$A_n(\mathbf{r}') = \frac{j\omega\mu d\phi_n^*(\mathbf{r}')}{k_n^2 - k^2} \tag{2.33}$$

Substituting A_n into Equation 2.30, the Green's function is obtained as

$$G(r,r') = j\omega\mu d \sum_n \frac{\phi_n(r)\phi_n^*(r')}{k_n^2 - k^2} \tag{2.34}$$

It should be noted that only open boundary is considered here to demonstrate the solution procedure. Equations 2.26 and 2.34 are valid for the PGPs with any boundary condition.

2.1.4 Eigenfunction for Power–Ground Planes with Rectangular Shape

v can be calculated by using the integral of the Green's functions, and the Green's function can be calculated by using the summation of eigenfunctions. Now, the problem is how to calculate the eigenfunction for a certain PGP. For the regularly shaped PGPs (such as rectangle, triangle, and circle), there is analytical solution of the eigenfunctions. For the arbitrarily shaped PGPs, numerical methods must be used to get the eigenfunctions. Since PCBs with rectangular shapes are commonly employed in practical designs, in this subsection, we will focus on the rectangular PGPs with the dimension $a*b$, as shown in Figure 2.3, to derive the eigenfunctions.

For PGPs, contours can be opened, shorted by shorting vias or decoupling capacitors, or partially opened and partially shorted. These different boundary conditions result in different formula of the eigenfunctions.

2.1.4.1 Eigenfunction for Open Boundary

Assume that the rectangular PGPs of Figure 2.3 have open boundaries. The corresponding eigenfunction can be obtained by using the separation of variables method. Let

$$\phi_n(r) = X(x)Y(y) \tag{2.35}$$

Figure 2.3 **Rectangular-shape PGPs (top view).**

Substituting Equation 2.35 into Equation 2.27, we obtain:

$$\frac{X''(x)}{X(x)} + \frac{Y''(y)}{Y(y)} + k_n^2 = 0 \qquad (2.36)$$

Let

$$\frac{X''(x)}{X(x)} = -k_{n_x}^2 \qquad (2.37)$$

$$\frac{Y''(y)}{Y(y)} = -k_{n_y}^2 \qquad (2.38)$$

where $k_{n_x}^2$ and $k_{n_y}^2$ are constants and $k_{n_x}^2 + k_{n_y}^2 = k_n^2$.
The solutions of $X(x)$ and $Y(y)$ are

$$X(x) = c_1 \cos(k_{n_x} x) + c_2 \sin(k_{n_x} x) \qquad (2.39)$$

$$Y(y) = c_3 \cos(k_{n_y} y) + c_4 \sin(k_{n_y} y) \qquad (2.40)$$

where c_1 to c_4 are unknown coefficients to be determined by using the open-boundary condition in Equation 2.28.

Applying the open-boundary condition at $x = 0$, we can get:

$$\left. \frac{\partial \phi_n(x, y)}{\partial x} \right|_{x=0} = k_{n_x} c_2 = 0 \qquad (2.41)$$

So, $c_2 = 0$ or $k_{n_x} = 0$; this results in $X(x) = c_1 \cos(k_{n_x} x)$. After that, we apply the open boundary condition at $x = a$ and get:

$$\left. \frac{\partial \phi_n(x, y)}{\partial x} \right|_{x=a} = -k_{n_x} c_1 \sin(k_{n_x} a) = 0 \qquad (2.42)$$

Since c_1 cannot be zero (otherwise, we have $\phi_n = 0$), we have $k_{n_x} = (n_x \pi / a)$, $n_x = 0,1, \ldots$.
Similarly, applying open-boundary conditions at $y = 0$ and $y = b$, we can get

$$C_4 = 0 \text{ and } \quad k_{n_y} = \frac{n_y \pi}{b}, \quad n_y = 0,1,\ldots \qquad (2.43)$$

Finally, the eigenfunction can be written as

$$\phi_n(\mathbf{r}) = c_5 \cos(k_{n_x} x) \cos(k_{n_y} y) \qquad (2.44)$$

where $c_5 = c_1 * c_3$. c_5 is determined by using the orthonormality of the eigenfunctions, as follow:

$$\int_S \phi_n(\mathbf{r}) \phi_m(\mathbf{r}) ds = c_5^2 \int_S \cos(k_{n_x} x) \cos(k_{m_x} x) \cos(k_{n_y} y) \cos(k_{m_y} y) ds$$

$$= c_5^2 \int_0^a \cos(k_{n_x} x) \cos(k_{m_x} x) dx \int_0^b \cos(k_{n_y} y) \cos(k_{m_y} y) dy$$

(2.45)

$$= \begin{cases} c_5^2 \dfrac{ab}{(2-\delta_{n_x})(2-\delta_{n_y})}, & m=n \\ 0, & m \neq n \end{cases} = \begin{cases} 1, & m=n \\ 0, & m \neq n \end{cases}$$

where δ_{nx} and δ_{ny} are the Kronecker delta with:

$$\delta_{n_x} = \begin{cases} 1, & \text{if } n_x = 0 \\ 0, & \text{if } n_x \neq 0 \end{cases}$$

(2.46)

$$\delta_{n_y} = \begin{cases} 1, & \text{if } n_y = 0 \\ 0, & \text{if } n_y \neq 0 \end{cases}$$

(2.47)

Therefore,

$$c_5 = \sqrt{\frac{(2-\delta_{n_x})(2-\delta_{n_y})}{ab}}$$

(2.48)

The eigenfunction can be written as

$$\phi_n(\mathbf{r}) = \sqrt{\frac{(2-\delta_{n_x})(2-\delta_{n_y})}{ab}} \cos(k_{n_x} x) \cos(k_{n_y} y)$$

(2.49)

The Green's function can therefore be written as

$$G(\mathbf{r},\mathbf{r}')$$

$$= \frac{j\omega\mu d}{ab} \sum_{n_x=0}^{\infty} \sum_{n_y=0}^{\infty} (2-\delta_{n_x})(2-\delta_{n_y}) \frac{\cos\left(\dfrac{n_x\pi}{a}x\right)\cos\left(\dfrac{n_y\pi}{b}y\right)\cos\left(\dfrac{n_x\pi}{a}x'\right)\cos\left(\dfrac{n_y\pi}{b}y'\right)}{\left(\dfrac{n_x\pi}{a}\right)^2 + \left(\dfrac{n_y\pi}{b}\right)^2 - k^2}$$

$$= \frac{j\omega\mu d}{ab} \sum_{m=0}^{\infty} \sum_{n=0}^{\infty} (2-\delta_m)(2-\delta_n) \frac{\cos(k_x x)\cos(k_y y)\cos(k_x x')\cos(k_y y')}{k_{mn}^2 - k^2}$$

(2.50)

with $k_{mn}^2 = k_x^2 + k_y^2$, $k_x = (m\pi/a)$, $k_y = (n\pi/b)$, $m, n = 0,1,2, \ldots$. Here and in the following sections, for simplicity, we use $m = n_x$ and $n = n_y$, and let ϕ_{mn} represent the eigenfunction with indices m and n.

2.1.4.2 Eigenfunctions for Other Boundaries

2.1.4.2.1 Eigenfunction for Short Boundary

When the contour of the top and bottom metal planes shown in Figure 2.3 is connected by shorting vias or decoupling capacitors, the boundary of the corresponding eigenfunctions can be considered as short boundary. We can follow the same procedure as mentioned in subsection 2.1.4.1 to get the eigenfunctions and the Green's function, where the boundary condition of the eigenfunction is changed to:

$$\phi_{mn}(x,y)\big|_{x=0} = 0 \tag{2.51}$$

$$\phi_{mn}(x,y)\big|_{x=a} = 0 \tag{2.52}$$

$$\phi_{mn}(x,y)\big|_{y=0} = 0 \tag{2.53}$$

$$\phi_{mn}(x,y)\big|_{y=b} = 0 \tag{2.54}$$

With this short boundary, the eigenfunction is obtained as

$$\phi_{mn}(x,y) = \frac{2}{\sqrt{ab}} \sin(k_x x)\sin(k_y y) \tag{2.55}$$

with $k_{mn}^2 = k_x^2 + k_y^2$, $k_x = (m\pi/a)$, $k_y = (n\pi/b)$, $m, n = 1,2,\ldots$

It should be noted that if $m = 0$ or $n = 0$, we will get $\phi_{mn} = 0$, so it is not included in the eigenfunction. The corresponding Green's function can be written as

$$G(\boldsymbol{r},\boldsymbol{r}') = j\frac{4\omega\mu d}{ab}\sum_{m=1}^{\infty}\sum_{n=1}^{\infty}\frac{\sin(k_x x)\sin(k_y y)\sin(k_x x')\sin(k_y y')}{k_{mn}^2 - k^2} \tag{2.56}$$

2.1.4.2.2 Eigenfunction for Hybrid Boundary

For some cases, the contour of the PGPs can be partially opened and partially shorted. One of such hybrid boundary, an example is shown in Figure 2.4, where the boundaries at $x = a$ and $y = 0$ are opened, but the boundaries at $x = 0$ and $y = b$ are shorted. For this hybrid boundary, the boundary condition of the eigenfunction is

$$\phi_{mn}(x,y)\big|_{x=0} = 0 \tag{2.57}$$

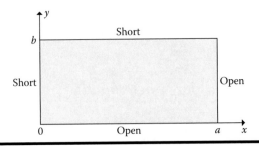

Figure 2.4 Hybrid boundary condition for PGPs (top view).

$$\left.\frac{\partial \phi_{mn}(x,y)}{\partial x}\right|_{x=a} = 0 \tag{2.58}$$

$$\left.\frac{\partial \phi_{mn}(x,y)}{\partial y}\right|_{y=0} = 0 \tag{2.59}$$

$$\left.\phi_{mn}(x,y)\right|_{y=b} = 0 \tag{2.60}$$

From Equations 2.57 and 2.58, we have:

$$X(x) = c_2 \sin(k_x x), \quad k_x = \frac{(m+1/2)\pi}{a}, \quad m = 0,1,2,\ldots \tag{2.61}$$

From Equations 2.59 and 2.60, we get:

$$Y(y) = c_3 \cos(k_y y), \quad k_y = \frac{(n+1/2)\pi}{b}, \quad n = 0,1,2,\ldots \tag{2.62}$$

According to the orthonormality, the eigenfunction is obtained as

$$\phi_{mn} = \frac{2}{\sqrt{ab}} \sin(k_x x)\cos(k_y y) \tag{2.63}$$

The corresponding Green's function is

$$G(\mathbf{r},\mathbf{r}') = j\frac{4\omega\mu d}{ab}\sum_{m=0}^{\infty}\sum_{n=0}^{\infty}\frac{\sin(k_x x)\cos(k_y y)\sin(k_x x')\cos(k_y y')}{k_{mn}^2 - k^2} \tag{2.64}$$

$$k_{mn}^2 = k_x^2 + k_y^2$$

2.2 Modal Field

2.2.1 Modal Field—From the View of Linear System

The eigenfunctions $\phi_{mn}(r)$ proposed in subsection 2.1.3.2 are also the mode functions of PGPs. They play an important role toward the electromagnetic property of PGPs.

Let us summarize the wave equation and its Green's function solution as follow:

$$(\nabla_t^2 + k^2)v(r) = -j\omega\mu d f(r) \tag{2.65}$$

$$v(r) = \int_S G(r,r')f(r')ds' \tag{2.66}$$

$$G(r,r') = j\omega\mu d \sum_{m,n} \frac{\phi_{mn}(r)\phi_{mn}^*(r')}{k_{mn}^2 - k^2}, \quad k_{mn}^2 = k_x^2 + k_y^2 \tag{2.67}$$

which is valid for arbitrarily shaped PGPs with any boundary condition. Since eigenfunctions form the basis functions, the excitation current $f(r)$ in Equation 2.65 can be expanded by using these eigenfunctions as

$$f(r) = \sum_{m,n} a_{mn}\phi_{mn}(r) \tag{2.68}$$

where a_{mn} is the expansion coefficients and $\phi_{mn}(r)$ in Equation 2.68 can be taken as the *excitation modes*.

According to Equations 2.66 and 2.67, and considering the orthogonality of the eigenfunctions in Equation 2.29, the voltage v can be calculated as

$$
\begin{aligned}
v(r) &= \int_S G(r,r')f(r')ds' \\[2mm]
&= \int_S j\omega\mu d \sum_{m,n} \frac{\phi_{mn}(r)\phi_{mn}^*(r')}{k_{mn}^2 - k^2} \cdot \sum_{m',n'} a_{m'n'}\phi_{m'n'}(r')ds' \\[2mm]
&= \sum_{m,n} \frac{j\omega\mu d}{k_{mn}^2 - k^2} a_{mn}\phi_{mn}(r) \\[2mm]
&= \sum_{m,n} H_{mn} a_{mn}\phi_{mn}(r)
\end{aligned}
\tag{2.69}
$$

where $H_{mn} = (j\omega\mu d / k_{mn}^2 - k^2)$, and $\phi_{mn}(r)$ in Equation 2.69 can be taken as the *response modes*.

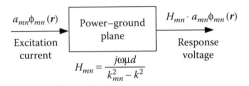

Figure 2.5 Response of PGPs to the excitation mode.

As shown in Figure 2.5, for each excitation mode, when it goes through PGPs, it will independently excite a response mode, and the ratio between the response mode and the excitation mode is H_{mn}. This mode analysis procedure is similar to the Fourier transform analysis of a linear system: the modes $\phi_{mn}(r)$ can be taken as the exponentials in Fourier transform; the indices m and n represent the different "frequency" components; and H_{mn} is the *impulse response* of the linear system (here, the PGPs).

From the impulse response H_{mn}, we can see that for a lossless PGPs system, it has infinite number of "self-excitation" when:

$$k^2 = \omega^2 \mu \varepsilon = k_x^2 + k_y^2 \tag{2.70}$$

These "self-excitations" denote the resonance of the PGPs, and the corresponding resonant frequencies can be derived from Equation 2.70 as

$$f_{mn} = \frac{c}{2\pi} \sqrt{k_x^2 + k_y^2} \tag{2.71}$$

where c is the speed of light in the substrate.

At or close to a certain resonant frequency f_{pq}, H_{pq} approaches infinity, so from Equation 2.69, the voltage distribution is totally decided by the mode function $\phi_{pq}(r)$, $v(r) \approx H_{pq} a_{pq} \phi_{pq}(r)$. The resonance is a big problem for PGPs, because it introduces signal integrity, power integrity, and EMI problems. In the following subsection, the mode function is studied in detail to better understand the behavior of the PGPs resonance.

According to the assumption in subsection 2.1.1, inside the substrate, the magnetic field is vertical to z direction, so the mode function $\phi_{pq}(r)$ is also called transverse magnetic (TM) (p, q) mode in some literatures.

2.2.2 Examples of Modal Fields

The first example is the PGPs with an open boundary, as shown in Figure 2.6. The substrate between the metal planes has a thickness of 1.2 mm, a relative permittivity of 4.4, and a loss tangent of 0.02. The resonant frequency of these PGPs is

$$f_{mn} = \frac{150}{\sqrt{\varepsilon_r \mu_r}} \sqrt{\left(\frac{m}{a}\right)^2 + \left(\frac{n}{b}\right)^2} \text{ GHz, with } m, n = 0, 1, 2, \dots \tag{2.72}$$

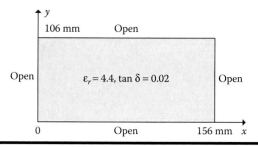

Figure 2.6   A power–ground plane with open boundary (top view).

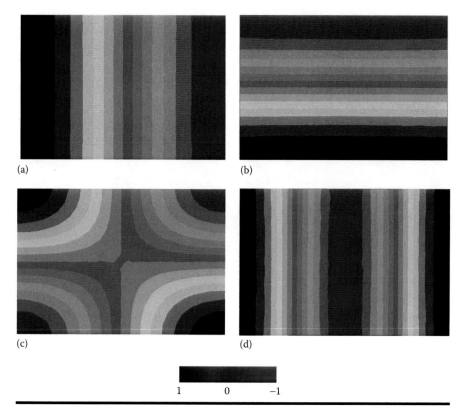

Figure 2.7   Normalized motions $\phi_{mn}(r)$ **for (a) (1,0) mode, (b) (0,1) mode, (c) (1,1) mode, and (d) (2,0) mode (top view).**

where a and b (in millimeter) are the length and width of the PGPs, respectively. ε_r and μ_r are the relative permittivity and permeability of the substrate, respectively.

For $a > b$, the lowest resonant frequency is $f_{10} = (150 / a\sqrt{\varepsilon_r \mu_r}) = 0.458\,\text{GHz}$, which is related to the dominant (1,0) mode of the PGPs. Figure 2.7a–d show the first four mode functions (the voltage distribution at resonant frequencies): ϕ_{10}, ϕ_{01},

Table 2.1   First Four Modes and Their Resonant Frequencies

Mode	Resonant Frequency (GHz)
(1,0)	0.458
(0,1)	0.675
(1,1)	0.815
(2,0)	0.916

ϕ_{11}, and ϕ_{20}. For the mode function $\phi_{mn}(r)$, the indices m and n represent the number of half-cycle voltage variations along the x and y directions, respectively. Table 2.1 lists the resonant frequencies of the first four modes.

The next example is the PGP shown in Figure 2.8. It has the same dimensions and substrate as those in Figure 2.6 but with a hybrid boundary. The resonant frequency of these PGPs is

$$f_{mn} = \frac{150}{\sqrt{\varepsilon_r \mu_r}} \sqrt{\left(\frac{m+0.5}{a}\right)^2 + \left(\frac{n+0.5}{b}\right)^2} \text{ GHz, with } m, n = 0,1,2,\ldots \qquad (2.73)$$

Its lowest resonant frequency is $f_{00} = 0.41$ GHz, which is related to the dominant (0,0) mode of the PGPs. Figure 2.8 also shows the mode function $\phi_{00}(r)$. We can see that by changing the boundary condition, the lowest resonant frequency and the mode function can be changed.

The third example is PGP with an arbitrary shape, as shown in Figure 2.9, where the substrate is the same as that in the above-mentioned examples. For this PGP, there is no analytical solution of the mode function. The mode function must be obtained by using numerical methods, such as the FEM and the method of moments. The dominant mode of this PGP is shown in Figure 2.9, where the resonant frequency is $f = 0.26$ GHz. The magnitude of the modal

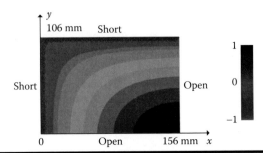

Figure 2.8   PGPs with hybrid boundary (top view) and its (0,0) mode.

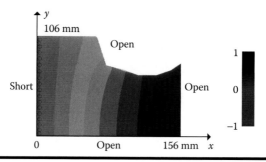

Figure 2.9   PGPs with an arbitrary shape (top view) and its dominant mode.

voltage achieve its maximum value at the open boundary and become zero at the short boundary, and the physical distance between the modal voltage maximum value and zero value decides the resonant wavelength and hence the resonant frequency of the mode.

2.2.3 Control of Modal Field

Besides the analysis of the modal field, the control of the modal field is important for PGPs design. For example, when the resonant frequency of the PGPs coincides with the working frequency, we need to shift it to avoid its interference, or when the sensitive signal trace passes through the PGPs, we need to change the modal field distribution to make a "clean room" for the through-hole via.

In recent years, the microstrip patch is widely employed in RF components' design, including antennas and filters, owing to its low cost and easy fabrication. Their structure is quite similar to the structure of PGPs. Most of these microstrip patches work at their resonant frequencies. The modal filed analysis and control mentioned in this section are also very important for their design. For example, the working bandwidth of a microstrip patch-based filter is usually narrow, with the center frequency satisfying $k = k_{mn}$ in Equation 2.67. In order to broaden its bandwidth, one way is to turn the modal fields distribution to make the resonant frequencies of two or more different modal fields much closer to each other, and then, the bandwidth of these modes will be connected together to form a wider bandwidth.

For a better design of these microstrip patch-based structure and components, we need some measures to control the modal fields of the PGPs structure. This can be easily done by adding shorting vias/decoupling capacitors or cutting slot on the PGPs with a low fabrication cost. The shorting via/decoupling capacitor changes the modal voltage, whereas the slot changes the modal current of the PGPs. Both of them can change the resonant frequencies. In the following subsection, the modal field and resonant frequencies after applying shorting vias/slot are obtained by using numerical methods.

2.2.3.1 Shorting Vias/Decoupling Capacitors

The shorting via connects the top and bottom metal planes, as shown in Figure 2.1 and then forces the voltage at that point to be zero. When it is applied at the location where the modal field magnitude has maximum values, it will greatly change the modal field distribution and then shift the corresponding resonant frequency. Usually, the resonant frequency will shift to a higher frequency by applying the shorting via. If the top and bottom metal planes are power and ground planes, respectively, decoupling capacitors, instead of shorting vias, are used to short the power and ground planes at high frequencies, while still keeping them isolated at DC.

The PGP shown in Figure 2.6 is discussed in this subsection to demonstrate the effect of the shorting vias on modal field and resonant frequencies. First, two shorting vias are applied at the centers of the left and right sides of the PGPs separately, as shown in Figure 2.10. From the original modal field distribution of Figure 2.7,

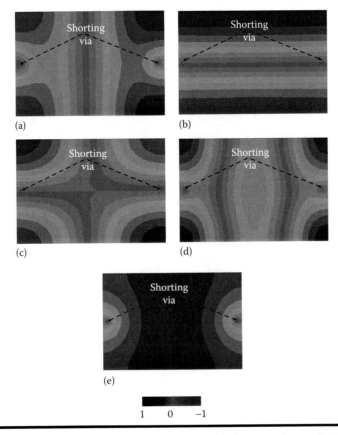

Figure 2.10 Normalized mode functions $\phi_{mn}(r)$ for (a) changed (1,0) mode, (b) (0,1) mode, (c) (1,1) mode, (d) changed (2,0) mode, and (e) additional mode (top view), after two shorting vias are applied.

Table 2.2 Comparison of Mode Resonant Frequencies

Mode	Resonant Frequency (GHz)	
	Without Shorting Vias	With Shorting Vias
(1,0)	0.458	0.656 ↑
(0,1)	0.675	0.670
(1,1)	0.815	0.815
(2,0)	0.916	1.024 ↑
Additional mode		0.306

we can see that the shorting vias are at the locations of the maximum magnitude of the original (1,0) and (2,0) modes, so they greatly change these two modal fields, as shown in Figure 2.10a and d. On the other hand, the shorting vias are at "zero" magnitude of the original (0,1) and (1,1) modes, so they almost have no effect on these two modal fields. The shorting vias also introduce an additional mode, as shown in Figure 2.10e. Table 2.2 lists the resonant frequencies of the modes with and without shorting vias. The table shows that after applying shorting vias, the resonant frequencies of the changed (1,0) and (2,0) modes shift to higher frequencies. At the same time, the resonant frequency of the (1,0) mode is much closer to that of the (0,1) mode. This is helpful in designing a filter with a wider bandwidth. The offset of the resonant frequencies can be controlled by adjusting the locations of the shorting vias.

Next, four shorting vias are applied on the four corners of the PGPs, as shown in Figure 2.11. For this time, the shorting vias are located at the maximum magnitude of all four modes, as shown in Figure 2.7, so all four modal fields are changed, as shown in Figure 2.11, and their resonant frequencies are pushed to higher frequencies. The changed resonant frequencies are listed in Table 2.3. Again, an additional mode appears, as shown in Figure 2.11e.

2.2.3.2 Slots

When we cut a slot on the top or bottom metal plane of Figure 2.1, it will change the surface current of the top or bottom metal plane. This results in the change of the modal field and resonant frequencies.

The PGP in Figure 2.6 is used again for the demonstration. A slot with dimensions 70 mm × 1 mm is cut on the center of the top metal plane, as shown in Figure 2.12, where the modal fields of the first four modes are also plotted. Table 2.4 lists the resonant frequencies with and without the slot. Since this slot is vertical to the surface current direction of the (1,0) mode (the surface current distribution can be found in Figure 2.15) and it locates at the maximum magnitude of the surface current, (1,0) modal field is greatly changed. The resonant frequency of the (1,0)

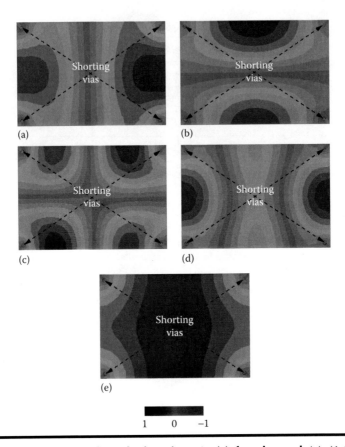

Figure 2.11 Normalized mode functions $\phi_{mn}(r)$ for changed (a) (1,0) mode, (b) (0,1) mode, (c) (1,1) mode, (d) (2,0) mode, and (e) additional mode (top view), after four shorting vias are applied.

Table 2.3   Mode Resonant Frequencies

	Resonant Frequency (GHz)	
Mode	*Without Shorting Vias*	*With Shorting Vias*
(1,0)	0.458	0.715 ↑
(0,1)	0.675	0.810 ↑
(1,1)	0.815	1.13 ↑
(2,0)	0.916	1.06 ↑
Additional mode		0.363

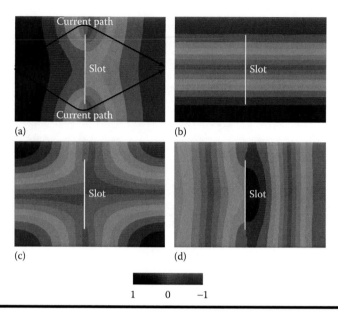

Figure 2.12 Normalized mode functions $\phi_{mn}(r)$ for (a) (1,0) mode, (b) (0,1) mode, (c) (1,1) mode, and (d) (2,0) mode (top view), after slot is applied.

Table 2.4 Comparison of Mode Resonant Frequencies

	Resonant Frequency (GHz)	
Mode	*Without Slot*	*With Slot*
(1,0)	0.458	0.370 ↓
(0,1)	0.675	0.673
(1,1)	0.815	0.810
(2,0)	0.916	0.905

mode also shifts to a lower frequency. This is because the slot increases the path of the (1,0) mode current, as shown in Figure 2.13. This resonant frequency offset can be easily controlled by adjusting the length and width of the slot.

Although the slot of the (2,0) mode is vertical to the surface current direction, it locates at the "zero" region of the current, as shown in Figure 2.15. Therefore, the resonant frequency shift of the (2,0) mode is not as large as that of the (1,0) mode. Similar explanation can be used for the (0,1) and (1,1) modes.

In this subsection, the control of modal field and resonant frequencies by using shorting via/decoupling capacitor and slot is discussed. These control methods are

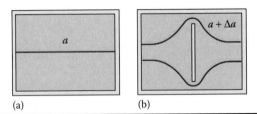

(a)   (b)

Figure 2.13   Current path of the (1,0) mode (a) without slot and (b) with slot.

intuitive and easy to understand. They are also flexible, since the resonant frequencies offset can be easily controlled by adjusting the location, number, and direction of the slots and vias.

2.2.4 *Induced Surface Current*

The induced surface current on the top and bottom metal planes of Figure 2.14 can be expressed as

$$J_S(r) = \begin{cases} -\hat{z} \times H(r), & z = d \\ \hat{z} \times H(r), & z = 0 \end{cases} \tag{2.74}$$

It should be noted that J_S is the induced current, which is different from the excitation current J_{Ext}. Since J_S on the top and bottom metal planes have the same distribution, in the following, we focus only on J_S on the bottom plane.

Substituting $J_S = \hat{z} \times H$ and $v = -E_z d$ into $\nabla \times E = -j\omega\mu H$, we get the relationship between J_S and v:

$$J_S = \frac{1}{-j\omega\mu} \hat{z} \times (\nabla \times E) = \frac{1}{j\omega\mu d} \hat{z} \times [\nabla \times (\hat{z}v)]$$

$$= \frac{1}{j\omega\mu d} \hat{z} \times (\nabla v \times z) = \frac{1}{j\omega\mu d} \nabla v \tag{2.75}$$

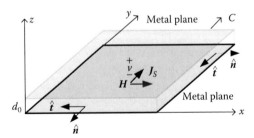

Figure 2.14   Surface current of PGPs.

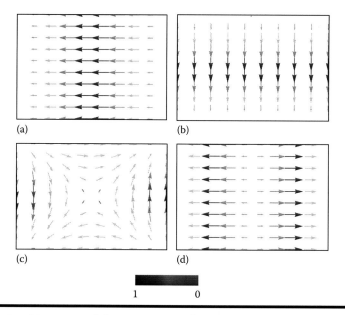

Figure 2.15 **Vector modal current** $\nabla\phi_{mn}$ **for (a) (1,0), (b) (0,1), (c) (1,1), and (d) (2,0) modes.**

Since v can be expressed by using the mode functions in Equation 2.69, J_S can also be expanded by using vector mode function $\nabla\phi_{mn}$. $\nabla\phi_{mn}$ is plotted in Figure 2.15 for the first four modes of the PGPs in Figure 2.7. The x-directional current distribution of (1,0) mode is equivalent to the current distribution of a half-wavelength dipole antenna. For the half-wavelength dipole antenna, its resonant frequency is inversely proportional to its length or current path length. Therefore, when the slot in Figure 2.12a increases the equivalent path length of the (1,0) modal current, its resonant frequency is reduced.

For the open boundary, along the contour C of the PGPs, we have:

$$0 = \hat{n} \times \mathbf{H} = -\hat{n} \times (\hat{z} \times J_S) = J_S(\hat{n} \cdot \hat{z}) - \hat{z}(\hat{n} \cdot J_S) = -\hat{z}(\hat{n} \cdot J_S) \qquad (2.76)$$

Therefore, current flowing outwardly from PGPs is zero, which agrees with the open boundary.

For the short boundary, along the contour C of the PGPs, we have:

$$0 = \hat{n} \cdot \mathbf{H} = -\hat{n} \cdot (\hat{z} \times J_S) = -\hat{t} \cdot J_S \qquad (2.77)$$

Therefore, $\boldsymbol{J}_S$ only have the component $J_{Sn}\hat{\boldsymbol{n}}$ normal to C. Substituting $\boldsymbol{J}_S = \hat{\boldsymbol{z}} \times \boldsymbol{H}$ into $\nabla \times \boldsymbol{H} = j\omega\tilde{\varepsilon}\boldsymbol{E}$ and considering $\hat{\boldsymbol{n}} \times \boldsymbol{E} = 0$ along C, we get:

$$0 = j\omega\tilde{\varepsilon}\hat{\boldsymbol{n}} \times \boldsymbol{E} = \hat{\boldsymbol{n}} \times (\nabla \times \boldsymbol{H}) = \hat{\boldsymbol{n}} \times [\nabla \times (\hat{\boldsymbol{z}} \times \boldsymbol{J}_S)] = \hat{\boldsymbol{n}} \times [\hat{\boldsymbol{z}}\nabla \cdot \boldsymbol{J}_S] = \hat{\boldsymbol{t}}\frac{\partial J_{Sn}}{\partial n} \tag{2.78}$$

Magnitude of J_{Sn} achieves its maximum value along C, which agrees with the short boundary.

2.3 Impedance Matrix of Power–Ground Planes

2.3.1 Port Definition

By using the Green's function, the impedance of the PGPs can be calculated. For this purpose, ports are defined at any location between the up and down planes, as shown in Figure 2.16. The port voltage and current are the voltage v and excitation current I_{Ext} in the wave Equation 2.15, respectively. The port current is the return current of the signal current flowing through the through-hole via inside the antipad. Figure 2.17 shows the relationship between the signal current, its return current, and the port voltage and current.

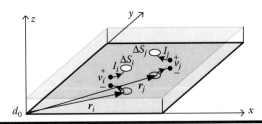

Figure 2.16 Ports' definition of the power–ground plane.

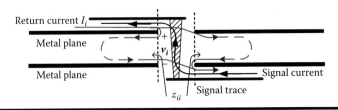

Figure 2.17 Signal current, its return current, and the port voltage and current at port *i* (cross section).

Assuming that I_j is applied on port j, the produced voltage at $\boldsymbol{r}$ is

$$v(\boldsymbol{r}) = \int_{\Delta S_j} G(\boldsymbol{r},\boldsymbol{r}')f(\boldsymbol{r}')ds' = \int_{\Delta S_j} G(\boldsymbol{r},\boldsymbol{r}')\frac{I_j}{\Delta s_j}ds' \approx I_j G(\boldsymbol{r},\boldsymbol{r}_j) \qquad (2.79)$$

where $\boldsymbol{r}_i$ and $\boldsymbol{r}_j$ are the centers of ports i and j, respectively, and ΔS_i and ΔS_j are the very small areas of ports i and j, respectively. The mutual impedance between ports i and j is

$$z_{ij} \equiv \frac{\dfrac{1}{\Delta s_i}\displaystyle\int_{\Delta S_i} v(\boldsymbol{r})ds}{I_j} \approx G(\boldsymbol{r}_i,\boldsymbol{r}_j) \qquad (2.80)$$

where v is averaged over the observation port i.

2.3.2 Equivalent Circuit

For rectangular PGPs with different boundary conditions, the mutual impedance z_{ij} can be written by using the eigenfunction, as

$$z_{ji} \approx \frac{j\omega\mu d}{ab}\sum_{m,n}\frac{\sqrt{ab}\phi_{mni}\sqrt{ab}\phi_{mnj}}{k_{mn}^2 - k^2} = \sum_{m,n}\frac{\sqrt{ab}\phi_{mni}\sqrt{ab}\phi_{mnj}}{\dfrac{k_{mn}^2 ab}{j\omega\mu d} + j\omega\tilde{\varepsilon}\dfrac{ab}{d}}$$

$$= \sum_{m,n}\frac{\sqrt{ab}\phi_{mni}\sqrt{ab}\phi_{mnj}}{\dfrac{1}{j\omega\dfrac{\mu d}{k_{mn}^2 ab}} + j\omega\varepsilon\dfrac{ab}{d} + \omega\varepsilon\dfrac{ab}{d}\tan\delta} \qquad (2.81)$$

where $k_{mn}^2 = k_x^2 + k_y^2$, $\phi_{mni} = \phi_{mn}(\boldsymbol{r}_i)$, and $\phi_{mnj} = \phi_{mn}(\boldsymbol{r}_j)$.

Defining the inductance $L_{mn} = (\mu d/k_{mn}^2 ab)$, capacitance $C_0 = \varepsilon(ab/d)$, conductance $G = \omega\varepsilon(ab/d)\tan\delta = \omega C_0 \tan\delta$, and their parallel circuit impedance $Z_{mn} = 1/(1/j\omega L_{mn}) + j\omega C_0 + G$, Equation 2.81 can be rewritten as

$$z_{ji} \approx \sum_{m,n}\frac{\sqrt{ab}\phi_{mni}\sqrt{ab}\phi_{mnj}}{\dfrac{1}{j\omega L_{mn}} + j\omega C_0 + G} = \sum_{m,n}\sqrt{ab}\phi_{mni}\sqrt{ab}\phi_{mnj}Z_{mn} \qquad (2.82)$$

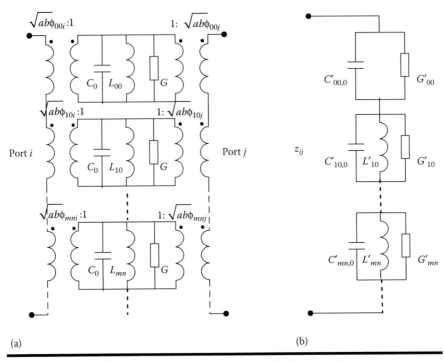

(a) (b)

Figure 2.18 **(a) Equivalent circuit of z_{ij} and (b) its simplified model, where** $C'_{mn,0} = (C_0/ab\phi_{mnj}\phi_{mnj})$, $L'_{mn} = ab\phi_{nni}\phi_{mnj}L_{mn}$, **and** $G'_{mn} = (G/ab\,\phi_{mni}\phi_{mnj})$.

z_{ij} is represented by using the equivalent circuit, as shown in Figure 2.18a. z_{ij} can be considered the series of infinite $C_0L_{mn}G$ parallel circuits. Each $C_0L_{mn}G$ parallel circuit denotes the resonance of the PGPs, the resonant frequency of which is f_{mn} in Equation 2.71. Another simplified form of equivalent circuit of z_{ij} is plotted in Figure 2.18b.

Before we calculate z_{ij}, let us explain more about the (0,0) mode in Equation 2.82. For the PGPs with an open boundary, from Equation 2.50, we can see that the contribution of the term with $m = n = 0$ to z_{ij} is

$$z_{ij}^{00}(r_i,r_j) = \frac{j\omega\mu d}{ab(-k^2)} = \frac{j\omega\mu d}{ab(-\omega^2\mu\varepsilon)} = \frac{1}{j\omega\dfrac{ab\varepsilon}{d}} = \frac{1}{j\omega C_0} \tag{2.83}$$

where we assume that the substrate is lossless. $C_0 = \varepsilon(ab/d)$ is just the DC capacitance between the up and down metal planes. $z_{ij}^{00}(r_i,r_j)$ is a constant, which is independent, with r_i and r_j. Therefore, a (0,0) mode stands for the DC uniform electric field distribution between up and down metal planes for the open boundary. If the boundary of the PGPs is short or partially short, its DC capacitance is zero. Therefore, it will not have a (0,0) mode (as for the short boundary in subsection 2.1.4.2.1) or its (0,0) mode does not contribute to the DC capacitance (as for the hybrid boundary in subsection 2.1.4.2.2). For the open boundary, $L_{00} \to \infty$ and can be removed from the equivalent circuit.

2.3.3 Characteristics of Impedance Curves

The PGP with an open boundary in Figure 2.6 is used here to analyze the characteristics of its impedance, where two ports are defined at (20 mm, 80 mm) and (117 mm, 80 mm), respectively, as shown in Figure 2.19. Figure 2.20 shows the calculated magnitudes of impedances, $|z_{11}|$, $|z_{12}|$, and $|z_{22}|$, which are obtained by using Equation 2.82. The characteristics of these impedances can be explained by the modes and equivalent circuit of the PGPs.

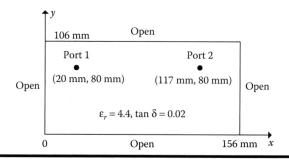

Figure 2.19 Power–ground plane with two ports defined, where the substrate thickness is 1.2 mm.

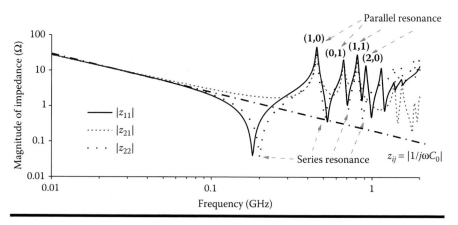

Figure 2.20 Magnitude of z_{ij}.

1. At low frequencies (from DC to the first-series resonant frequency), due to the size of the PGP being smaller than the working wavelength, it can be taken as a parallel plate capacitor with capacitance C_0. The value of C_0 can be calculated by the slope of impedances, as shown in Figure 2.20, and it is independent of the location of ports.

2. At high frequencies, the PGPs have infinite resonant modes. These resonant modes result in the impedance peaks in Figure 2.20, such as the (1,0) to (2,0) modes of Figure 2.7. At its resonant frequency f_{mn}, the impedance of $C_0 L_{mn} G$ parallel circuit becomes much larger, which prevents the current flowing at ports i and j in Figure 2.18.

3. At high frequencies, there are also local minima on the impedance curves. For a parallel LC circuit, it can be taken as a frequency-dependent inductor for frequencies below its resonant frequency, since the imaginary part of its admittance is negative, whereas it can be taken as a frequency-dependent capacitor for frequencies above its resonant frequency. According to this, for a lossless substrate, when the frequency is between two mode resonant frequencies, for example, f_{mn} and f_{m+1n}, the equivalent circuit of z_{ii} can be reduced to the series of C_Σ and L_Σ, as shown in Figure 2.21. This $C_\Sigma L_\Sigma$ series circuit results in the local minima at its resonant frequency $(1/2\pi\sqrt{C_\Sigma L_\Sigma})$.

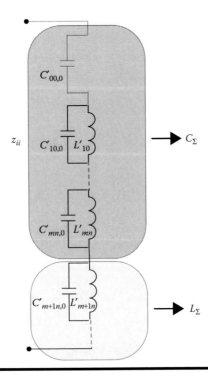

Figure 2.21   Reduced equivalent circuit of z_{ii}.

Considering the loss of substrate, the local minima in Figure 2.20 are not zero. It is due to the same reason as for the mutual impedance local minima. For clarity, in Figure 2.20, we refer the mode resonances as the *parallel resonances* and refer these $C_\Sigma L_\Sigma$ resonances as the *series resonances*.

4. All of the resonant frequencies related to four modes in Figure 2.7 are observed on $|z_{11}|$ curve; however, f_{20} is not observed on $|z_{12}|$ and $|z_{22}|$ curves. This is because port 2 is located at the zero of the (2,0) mode, which results in $\phi_{20j} = 0$ in equivalent circuit of Figure 2.18. It should be noted that the resonant modes are independent of the excitation and observation ports; however, whether these resonant modes can be "seen" from the impedance is decided by the ports' locations.

The above-mentiones conclusions are also valid for the PGPs with arbitrary shapes and any boundary; their impedance curves can always be divided into three different regions: the low-frequency capacitance (for open boundary), the high-frequency parallel resonance, and the high-frequency series resonance regions.

2.3.4 Equivalent Network

After ports are defined, the PGPs can be represented by using an equivalent network, the impedance matrix of which can be obtained by Equation 2.82 for rectangular shapes or by numerical methods for arbitrary shapes. This equivalent network can then be connected with equivalent networks of other components and circuits to perform the whole system simulation.

For PGPs with two ports, shown in Figure 2.19, two decoupling capacitors are applied at the centers of the left and right sides of the PGPs separately (as shown in Figure 2.10, where the shorting vias are replaced by decoupling capacitors). The whole structure can be equivalent to a four-ports network, as shown in Figure 2.22, where port 3 and port 4 are terminated with the decoupling capacitors. The decoupling capacitor is 200 pF, with an equivalent series inductance (ESL) of 0.01 nH.

The impedance matrix $[Z]$ of this four-port network is obtained by Equation 2.82, and then, $|z_{11}|$ is calculated and shown in Figure 2.23, where, for

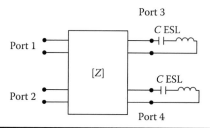

Figure 2.22 Four-port network of the power–ground plane.

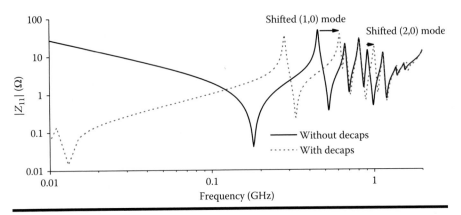

Figure 2.23 $|z_{11}|$ **with and without the decoupling capacitors.**

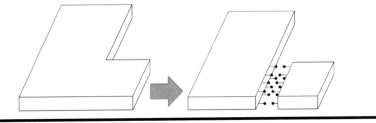

Figure 2.24   Segment method.

comparison, $|z_{11}|$ without the decoupling capacitors is also plotted. From this figure, we can see that after applying the decoupling capacitors, the resonant frequencies of the (1,0) and (2,0) modes shifted to higher frequencies, which agree with the results in Table 2.2.

The ports can also be defined along the contour of the PGPs. In this case, the PGPs with arbitrary shapes can be solved by using the Green's function of PGPs with rectangular shape. This is also called the *segment method*, as shown in Figure 2.24. In Figure 2.24, the *L*-shaped plane is divided into two rectangular planes and connecting ports are defined along their interface.

2.4 Imaging Method

In the earlier sections, we had discussed the mode analysis of the PGPs. Owing to its intuitive nature, the mode expansion method is widely used in real engineering applications, such as the electromagnetic wave propagation analysis of reverberation chamber and rectangular cavity [12,13]. The pair of PGPs forms a cavity at high frequency. Lots of electromagnetic modes exist inside the cavity. The Green's function must consider contributions from all those modes to get an accurate result when it is applied to calculate the impedance of PGPs at a high frequency. This makes the mode expansion method inefficient at a high frequency. In this section, an alternative analytical method, imaging method, is proposed to calculate the Green's function of the PGPs structure [7]. It can be found that the imaging expansion method can achieve faster convergence at a high frequency than the mode expansion method. However, the imaging method shows a slower convergence than the mode expansion method at a low frequency. In fact, the mode expansion method and the imaging method are equivalent when we consider infinite number of modes and images in their expression formulas. Based on this equivalence, we propose a hybrid method to take the advantages of both mode expansion method and imaging method.

2.4.1 Problem Statement

The PGP with a rectangular shape under study in this subsection is shown in Figure 2.25, which consists of two metal planes with a length of a and a width of b. The substrate sandwiched between these two metal planes has a relative permittivity of ε_r, a loss tangent of $\tan\delta$, and a thickness of d. Two ports with a radius of R are located at the observation point $r = x\hat{x} + y\hat{y}$ and the source point $r' = x'\hat{x} + y'\hat{y}$, respectively.

Previously, we had got the Green's function of the PGPs in Figure 2.25, which is expressed by using the mode functions. In order to distinguish it from the Green's functions obtained from images, we refer to it as G_{mode} and rewrite it as

$$G_{\text{mode}}(r,r')$$

$$= \frac{1}{ab} \sum_{m=0}^{\infty} \sum_{n=0}^{\infty} (2-\delta_m)(2-\delta_n) \frac{\cos(k_x x)\cos(k_y y)\cos(k_x x')\cos(k_y y')}{k_{mn}^2 - k^2} \qquad (2.84)$$

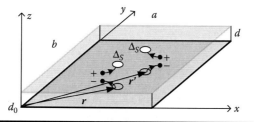

Figure 2.25 Power–ground plane with two ports.

with $k_{mn}^2 = k_x^2 + k_y^2$, $k_x = (m\pi/a)$, $k_y = (n\pi/b)$, $m, n = 0,1,2,....$ It should be noted that $j\omega\mu d$ is removed from G_{mode}. When frequency becomes higher, the number of modes also increases, so that the convergence of G_{mode} becomes very slow.

2.4.2 Imaging Method

The imaging method is postulated as an alternative solution for the Green's function. Figure 2.26 shows one example of the imaging method for a pair of rectangular PGPs. Taking the four PMC boundaries of the PGPs in consideration, we use multi-images of the source port to represent the wave reflections from the PMC boundary. In that case, the PMC boundary can be removed and the electromagnetic field arriving at the observation port can be taken as the radiation contribution from the source port and all its images. For the PMC boundary, the source and its images will have the same value and polarization, which are shown in Figure 2.26. The position of the image can be derived as

$$r'_{mn} = (\alpha x' - 2ma)\hat{x} + (\beta y' - 2nb)\hat{y} \tag{2.85}$$

with $\alpha/\beta = -1$ or 1 and $m/n = -\infty...\infty$. For each (m,n) pair, there are four images. Since the PMC boundary is now replaced with the virtual images of the source via, the Green's function of 2D free space can be used. Considering the free-space radiation boundary condition, together with the wave equation, the Green's function at r due to the source at r' can be derived as

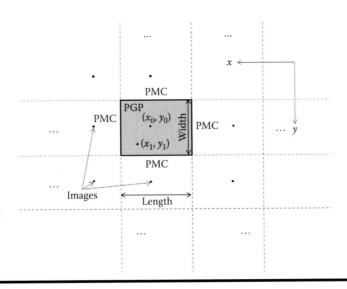

Figure 2.26   Images of the source.

$$G(r,r') = \frac{1}{4j} H_0^{(2)}(k\rho) \tag{2.86}$$

where $\rho = |r - r'|$ and $k = \omega\sqrt{\varepsilon\mu}$ is the wave number. $H_0^{(2)}(k\rho)$ is a zero-order Hankel function of the second kind. In case the imaginary part of k is negative, $H_0^{(1)}(k\rho)$ is used instead of $H_0^{(2)}(k\rho)$.

The final imaging Green's function can be written by adding the contributions from the source and all its images, and this Green's function is referred to as G_{image}

$$G_{\text{image}} = \frac{1}{4j} \sum_{m,n=-\infty}^{\infty} \sum_{\alpha,\beta} H_0^{(2)}(k\rho_{mn,\alpha\beta}) \tag{2.87}$$

where $\rho_{mn,\alpha\beta} = \sqrt{(X_\alpha + 2ma)^2 + (Y_\beta + 2nb)^2}$, $X_\alpha = x - \alpha x'$, $Y_\beta = y - \beta y'$, and $k = \omega\sqrt{\mu\varepsilon}$.

For $r = r'$, $\rho_{mn,\alpha\beta}$ with $m = n = 0$ and $\alpha = \beta = 1$ will become zero, which gives the singularity of the Green's function. For this case, the term $(1/4j) H_0^{(2)}(k\rho_{00,11})$ can be averaged over the area of the port as: $(1/\pi R^2)\int_{\Delta s} (1/4j) H_0^{(2)}(k\rho_{00,11}) ds' \approx (1/4\pi)[1 - 2\ln(kR)]$, where R is the radius of the port.

When $\rho_{mn,\alpha\beta}$ increases, the Hankel functions decay quickly; therefore, we just need to calculate a few images that are near to the observation port to get accurate mutual impedance.

2.4.3 Hybrid Method

In the following subsection, the Ewald identity is employed to build the link between the mode Green's function G_{mode} and the imaging Green's function G_{image}. Furthermore, we proposed a hybrid imaging and mode method to solve the PGPs problem.

The 2D Ewald identity of Hankel function can be written as

$$\frac{1}{4j} H_0^{(2)}(kR) = \frac{1}{2\pi} \int_0^{\infty} \frac{e^{-R^2 s^2 + \frac{k^2}{4s^2}}}{s} ds \tag{2.88}$$

By substituting Equation 2.88 into Equation 2.87, we can rewrite the imaging Green's function as

$$G_{\text{image}} = G_1 + G_2 \tag{2.89}$$

where

$$G_1 = \frac{1}{2\pi} \sum_{\alpha,\beta} \sum_{m,n=-\infty}^{\infty} \int_0^E \frac{e^{-\rho_{mn,\alpha\beta}^2 s^2 + \frac{k^2}{4s^2}}}{s} ds \tag{2.90}$$

$$G_2 = \frac{1}{2\pi} \sum_{\alpha,\beta} \sum_{m,n=-\infty}^{\infty} \int_E^{\infty} \frac{e^{-\rho_{mn,\alpha\beta}^2 s^2 + \frac{k^2}{4s^2}}}{s} \, ds \qquad (2.91)$$

Factor E is a splitting parameter. Next, we expand G_1 as follows:

$$G_1 = \frac{1}{2\pi} \sum_{\alpha,\beta} \int_0^E \frac{1}{s} e^{\frac{k^2}{4s^2}} \sum_{m=-\infty}^{\infty} e^{-\left(\frac{X_\alpha}{2a}+m\right)^2 (2as)^2} \sum_{n=-\infty}^{\infty} e^{-\left(\frac{Y_\beta}{2b}+n\right)^2 (2bs)^2} \, ds \qquad (2.92)$$

By using Poisson summation formula [14], we can get:

$$\sum_{l=-\infty}^{\infty} e^{-(l+g)^2 s^2} = \frac{\sqrt{\pi}}{s} \sum_{l=-\infty}^{\infty} e^{-\left(\frac{l\pi}{s}\right)^2 - j 2\pi l g} \qquad (2.93)$$

By replacing variable $l + g$ with $(X_\alpha/2a) + m$ and $(Y_\beta/2b) + n$, respectively, in Equation 2.93 and substituting Equation 2.93 into Equation 2.92, we get:

$$G_1 = \frac{1}{8ab} \sum_{\alpha,\beta} \int_0^E \frac{e^{\frac{k^2}{4s^2}}}{s^3} \sum_{m=-\infty}^{\infty} e^{-\left(\frac{m\pi}{a}\right)^2 \frac{1}{4s^2}} e^{-j\frac{m\pi}{a} X_\alpha} \sum_{n=-\infty}^{\infty} e^{-\left(\frac{n\pi}{b}\right)^2 \frac{1}{4s^2}} e^{-j\frac{n\pi}{b} Y_\beta} \, ds \qquad (2.94)$$

The following integration can be calculated as

$$\int_0^E \frac{1}{s^3} e^{-\frac{\kappa_{mn}^2}{4s^2}} \, ds = \frac{-4}{\kappa_{mn}^2} \cdot \frac{1}{-2} \int_0^E de^{-\frac{\kappa_{mn}^2}{4s^2}} = \frac{2e^{-\frac{\kappa_{mn}^2}{4E^2}}}{\kappa_{mn}^2} \qquad (2.95)$$

After that, we define $\kappa_{mn}^2 = k_x^2 + k_y^2 - k^2$ with $k_x = m\pi/a$ and $k_y = n\pi/b$; for $m, n = 0,1,2,\ldots\infty$, $C_{X_\alpha} = \cos(k_x X_\alpha)$ and $C_{Y_\beta} = \cos(k_y Y_\beta)$. Substituting Equation 2.95 into Equation 2.94, G_1 can be written as

$$G_1 = \frac{1}{4ab} \sum_{\alpha,\beta} \sum_{m,n=0}^{\infty} (2 - \delta_m)(2 - \delta_n) C_{X_\alpha} C_{Y_\beta} \frac{e^{-\frac{\kappa_{mnl}^2}{4E^2}}}{\kappa_{mnl}^2}$$

$$(2.96)$$

$$= \frac{1}{4ab} \sum_{m,n=0}^{\infty} (2 - \delta_m)(2 - \delta_n) \frac{e^{-\frac{\kappa_{mn}^2}{4E^2}}}{\kappa_{mn}^2} \sum_{\alpha} C_{X_\alpha} \sum_{\beta} C_{Y_\beta}$$

where

$$\sum_{\alpha} C_{X\alpha} = \cos[k_x(x-x')] + \cos[k_x(x+x')] = 2\cos(k_x x)\cos(k_x x') = 2C_x C_{x'} \quad (2.97)$$

Finally,

$$G_1 = \frac{1}{ab} \sum_{m,n=0}^{\infty} \frac{(2-\delta_m)(2-\delta_n)}{\kappa_{mn}^2} e^{-\frac{\kappa_{mn}^2}{4E^2}}$$

$$\cos\left(\frac{m\pi}{a}x\right)\cos\left(\frac{m\pi}{a}x'\right)\cos\left(\frac{n\pi}{b}y\right)\cos\left(\frac{n\pi}{b}y'\right) \quad (2.98)$$

According to [15], we can get:

$$G_2 = \frac{1}{4\pi} \sum_{m,n=-\infty}^{\infty} \sum_{\alpha,\beta} \sum_{q=0}^{\infty} \left(\frac{k}{2E}\right)^{2q} \frac{1}{q!} E_{q+1}(\rho_{mn,\alpha\beta}^2 E^2) \quad (2.99)$$

where E_q is a q-order exponential integral function, $E_q(z) = \int_1^{\infty}(e^{-zt}/t^q)dt$. $E_q(z)$ decays as e^{-z}/z as $|z| \rightarrow \infty$.

Let $E \rightarrow \infty$, from Equation 2.91, we can see that $G_2 = 0$, whereas $e^{-\kappa_{mn}^2/4E^2} \rightarrow 1$ in Equation 2.98. Therefore, G_{image} is reduced to G_{mode}. This verifies that the Green's functions obtained from mode expansion and images are equivalent to each other.

In real application, we must truncate the infinite number of m and n and neglect the higher-order terms with larger m and n. This results in computation error and different convergences of the mode and imaging methods at different frequencies. Mode method converges quickly at low frequencies, whereas imaging method converges quickly at high frequencies. From Equation 2.89, we can take a proper E factor to propose a new hybrid method, which takes advantages of both imaging and mode methods. If we choose a larger/smaller E, the hybrid method will be reduced to mode/image method. In Equation 2.98, m and n can be considered the mode number of hybrid method, like that in the mode method. In addition, in Equation 2.99, $\rho_{mn,\alpha\beta} = \sqrt{(X_\alpha + 2ml)^2 + (Y_\beta + 2nw)^2}$ is the distance from the images to the observation port.

2.4.4 Validation

In this subsection, we examine the accuracy and efficiency of the proposed hybrid method, where the calculated impedance of PGPs by using the hybrid method, imaging method, mode method, and full-wave method as well as their measurement results are compared. The equivalence between the hybrid, imaging, and mode methods is validated. The results show that the hybrid method has a faster convergence than the mode/imaging method in higher-/lower-frequency band.

2.4.4.1 Validation of the Hybrid Method

The first example is a PGP structure with two vias, as shown in Figure 2.27, where the mutual impedance between two through-hole vias is calculated. The dielectric material is FR4 with a relative permittivity of 4.4 and a loss tangent of 0.02. The board has a length of 100.15 mm, a width of 75.10 mm, and a height of 1.65 mm. Two ports are defined at two through-hole vias at (30 mm, 30 mm) and (70 mm, 55 mm), respectively. The fabricated PCB is shown in Figure 2.27.

The S parameters are simulated and compared. The amplitudes of the S parameters are calculated by using the proposed hybrid method with $4 \times 11 \times 11$ images and 900 modes, which are compared with the results from experiment from 0.4 to 5.0 GHz, as shown in Figure 2.28. Good agreement is observed from 0.4 to 3 GHz.

Figure 2.27   PCB test board for the validation.

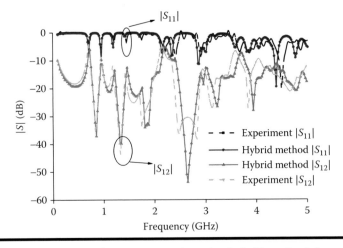

Figure 2.28   Calculation of S parameters by using hybrid method, in comparison with experimental data for frequency band (0.4–5 GHz).

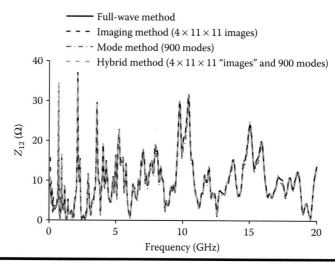

Figure 2.29 Z_{21} **calculation using the mode, imaging, full-wave, and hybrid methods in 0.1–20 GHz.**

Next, the hybrid method is used to calculate the Z parameter, the result of which is compared with those of the mode, imaging, and full-wave methods. The dimensions and materials of the PGPs are the same as those in the first example. Z_{12} is shown in Figure 2.29, where the number of modes and images used in the three methods (the mode method, the imaging method, and the hybrid method) are listed in Figure 2.29.

Figure 2.29 shows that the results generated by the four methods agree well, but the time consumption is different. The three analytical methods use much less time than the full-wave method (seconds vs. several tens of minutes). Thus, we can see that the hybrid method is more efficient than the full-wave method.

2.4.4.2 Comparison of Convergence of Three Analytical Methods at Low Frequency

In the third example, the hybrid method is used to calculate the Z parameter and is compared with the mode, imaging, and full-wave methods at low frequency. The dimensions and materials of the PGPs are the same as those in the first example.

Figure 2.30 exhibits that the proposed hybrid method generates the same result as that of the mode method. Their results agree well with those of the full-wave method and are more accurate than those of the imaging method. Actually, when we decrease the number of images (from $4 \times 11 \times 11$ to $4 \times 7 \times 7$) of the hybrid method, the result does not change too much, as shown in Figure 2.7. This shows that the main determining factor of the hybrid method at low frequency is the mode part.

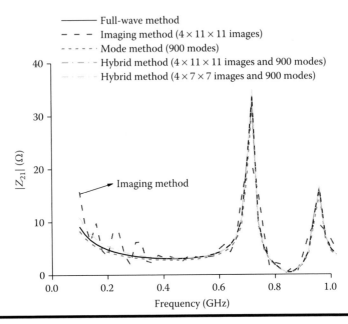

Figure 2.30 Z_{21} **calculation using the mode, imaging, full-wave, and hybrid methods in 0.1–1 GHz.**

The hybrid method fits well with the full-wave method and is better than the imaging method. The imaging method does not get good convergence, because at low frequency, the wave number k in Equation 2.87 is small and the images of source at far region will still contribute to E-field at the victim via. Therefore, more images need to be calculated to make the imaging method converged. The hybrid method is more efficient and accurate than the imaging method at low frequency.

2.4.4.3 Comparison of Convergence of Three Analytical Methods at High Frequency

Figure 2.31a,b show the calculated Z_{21} at high-frequency range by using four methods: the full-wave method, the mode method, the imaging method, and the proposed hybrid method. From Figure 2.31, a good agreement between the proposed hybrid method, the imaging method, and the full-wave method can be observed. The mode method shows a little error in comparison with the full-wave method. This is due to the non-convergence of the mode method at high frequency. In order to get a more accurate result, more higher-order modes should be considered at a high frequency than that at a low frequency. However, the hybrid method calculates the contribution from those higher-order modes by considering Equation 2.99, so that it can get a more accurate result at high frequency than the mode method does. When we reduce the number of mode used in the hybrid method (from 900

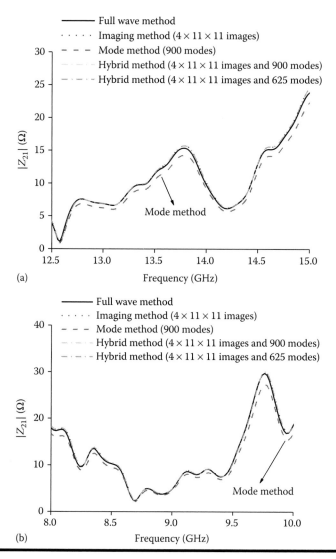

Figure 2.31 **(a)** Z_{21} **calculation using the mode, imaging, full-wave, and hybrid methods in 8–10 GHz. (b)** Z_{21} **calculation using the mode, imaging, full-wave, and hybrid methods in 12.5–15 GHz.**

to 625), its result in Figure 2.31 still shows a good agreement with the full-wave method. This implies that the imaging part in the hybrid method is the major contributor at high frequency.

Figure 2.32 shows the calculated Z_{11} at high frequency. Good agreement between the proposed hybrid method, the imaging method, and the full-wave method can also be observed. The mode method shows worse discrepancy in comparison with

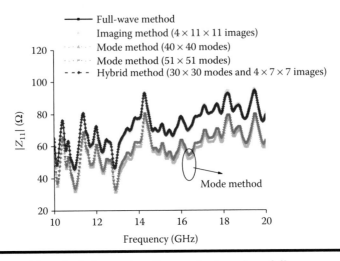

Figure 2.32 Z_{11} **calculation using the mode, imaging, full-wave, and hybrid methods in 10–20 GHz.**

the full-wave method, even when we use more modes. The reason is that when two vias get closer, there are more higher-order interactions of modes between the two vias; therefore, we need to calculate more modes to get accurate results. The self-impedance can be considered as two vias at same position, and it is much harder to get convergence when calculating Z_{11} by using mode method. However, the hybrid method can get convergence easily.

2.4.4.4 Comparison of Computing Time

Table 2.5 shows the time consumption of all four methods. The computer has a CPU of four cores at 2.8 GHz and 8 GB RAM. The remarks state the condition under which the calculations are carried out.

Table 2.5 illustrates that the computing time by three analytical methods is in the range of a few seconds compared with the computing time by the full-wave method, which is in the range of minutes. In order to get good accuracy at both low and high frequencies, the number of modes for the mode method and the number of images for the imaging method must be large enough (in this case, there should be 10,000 modes and 4 × 41 × 41 or more images). The proposed hybrid method is an adaptive method and can be automatically reduced to the mode method and the imaging method at low and high frequencies, respectively. Therefore, we can conclude that the proposed hybrid method is the best trade-off between the computational accuracy and efficiency, covering both low and high frequency range.

Table 2.5 Time Consumption of Calculation Using the Mode, Imaging, Full-wave, and Hybrid Methods at 0.1–20 GHz

Method	Time (seconds)	Remarks
Full-wave	4,113	0.1 GHz–20 GHz 996 points 6,777 tetrahedra
Mode method	1.055	900 modes
	1.801	1600 modes
	2.722	2,500 modes
	10.26	10,000 modes
Imaging method	1.584	4 × 11 × 11 images
	2.996	4 × 21 × 21 images
	5.946	4 × 41 × 41 images
Hybrid method	3.261	4 × 7 × 7 images and 900 modes
	4.393	4 × 7 × 7 images and 1,600 modes
	5.926	4 × 11 × 11 images and 900 modes

2.5 Power–Ground Grids

With the development of 2.5D integration technology, the interposer technology is widely used to integrate different ICs into one electronic package. Since a solid PGP is easy to peel off during the fabrication process, PDN on interposer is usually designed to be a grid. Several previous works have been carried out to model such PGG on interposer. In [16], a segmentation method was proposed for modeling large-sized PDN on silicon interposer, where the whole PGG was divided into many small segments and the impedance was obtained by solving a complex equivalent circuit. Because of its complex structure, a 3D full-wave simulator has been used to estimate such PGG impedance, which can obtain accurate results. Unfortunately, the PGG on interposer is composed of quantities of vias and it calls for a large number of meshing, which results in a long simulation time as well as requirement for large computing resources. Therefore, it is necessary to develop an efficient and accurate method to calculate the impedance of the PGG on interposer.

In this section, a semi-analytical method for calculating the impedance of the PGG on interposer is proposed. An equivalent PGP for the PGG is constructed. First, both PGG and its equivalent PGP are divided into many unit cells, where resistance (R), inductance (L), capacitance (C), and conductance (G) of each unit cell of PGG are extracted. Then, the unit cell of equivalent planes can be constructed, which has the same *RLCG* values as that of the grid-type unit cell. Since the conductive loss and dielectric loss only affect the Q factor, instead of the resonant frequency of the impedance curve, the metal layers are assumed to be PEC and the medium is assumed to be lossless. The equivalent-planes unit cell has the same size as that of the grid-type unit cell, but the medium is different. Therefore, the entire equivalent PGP is obtained by combining the equivalent-planes unit cells together, and the impedance of the equivalent PGP (and also the PGG) can be analytically calculated by using the resonant cavity method or the imaging method discussed in the previous sections. To further improve the accuracy of the proposed formula, an empirical formula is proposed to modify the resonant cavity mode for the equivalent PGP to ensure that it can obtain consistent result with the PGG. The proposed method is validated by the full-wave simulation method and is compared with an available measurement result.

2.5.1 Equivalent Power–Ground Plane of the Power–Ground Grid

The structure of a typical PGG on interposer is shown in Figure 2.33. It is composed of two metal layers with a length of *a* and a width of *b*. The light lines of the structure represent the power grids, and the dark ones are the ground grids. Through vias are used to connect the PGGs from the top layer to the bottom layer. This PGG can take a periodic structure, and Figure 2.34a represents its unit cell. L_w is the line width of the metal layer, and L_p is the pith between grids. The position of the connecting vias is also displayed in the figure, with a diameter of d_{via}. The side view of this unit cell is shown in Figure 2.34b, where *t* is the thickness of the metal layer and *d* is the thickness of the substrate sandwiched between the

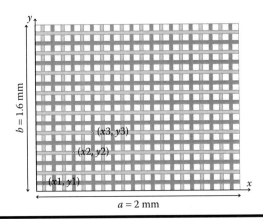

Figure 2.33 Top view of the geometry of the power–ground grid.

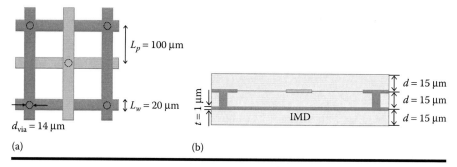

Figure 2.34 **Unit cell of the power–ground grid. (a) Top view and (b) side view.**

two metal layers. In our simulation, the substrate is set as an intermetal dielectric (IMD) with a relative permittivity of ε_{IMD}.

The grid-type unit cell used to extract the inductance (L_g) and capacitance (C_g) is shown in Figure 2.35a, where L_1 represents the self-inductance of the power grid, whereas L_2 is the self-inductance of the ground grid. M is the mutual inductance between the power grid and the ground grid. The two current paths from source to sink in Figure 2.35a are set up to obtain L_1 and L_2 by using numerical quasi-static methods. The final inductance of the grid-type unit cell can be determined by

$$L_g = L_1 + L_2 - 2M \tag{2.100}$$

The parameter, C_g, is the capacitance of the grid-type unit cell between the ground grid and the power grid, as shown in Figure 2.35a, which is also extracted directly by the quasi-static method. The equivalent-planes unit cell is constructed with the same L_g and C_g values. The structure of the equivalent-planes unit cell is show in Figure 2.35b. It is assumed that the size of the equivalent-planes unit cell is the same as that of the grid-type unit cell. Therefore, the difference between two structures is a characteristic of the substrate.

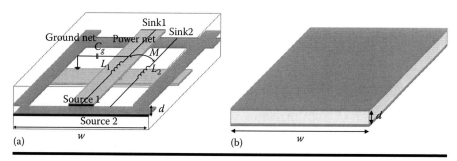

Figure 2.35 **(a) Unit cell of the power–ground grid for extracting the inductance and capacitance and (b) the constructed equivalent-planes unit cell.**

For the equivalent-planes unit cell, we have:

$$L_g = \mu_g d \tag{2.101}$$

$$C_g = \varepsilon_0 \varepsilon_g \frac{w^2}{d} \tag{2.102}$$

where μ_g and ε_g are the permeability and relative permittivity of the substrate of the equivalent-planes unit cell, respectively. ε_0 is the permittivity of free space; w and d are the length of the equivalent planes unit cell (and also the grid-type unit cell) and the thickness between two metal layers, respectively.

With Equations 2.101 and 2.102, we can derive the permeability and relative permittivity of the substrate of the equivalent-planes unit cell as

$$\mu_g = \frac{L_g}{d} \tag{2.103}$$

$$\varepsilon_g = C_g \frac{d}{\varepsilon_0 w^2} \tag{2.104}$$

The constructed equivalent-planes unit cell has the same inductance and capacitance as that of the grid-type unit cell. The dimension of the entire equivalent PGP remains the same as that of the PGG with a length of a and a width of b. The impedance of the original PGG can be calculated by using its equivalent PGP as:

$$z_{ji} \approx \frac{j\omega\mu_g d}{ab} \sum_{m,n} \frac{\sqrt{ab}\phi_{mni}\sqrt{ab}\phi_{mnj}}{k_{mn}^2 - k^2} \tag{2.105}$$

where $k_{mn}^2 = k_x^2 + k_y^2$, $\phi_{mni} = \phi_{mn}(r_i)$, $\phi_{mnj} = \phi_{mn}(r_j)$, and $k = \omega\sqrt{\varepsilon_g\mu_g}$.

To verify the accuracy of the equivalent-planes representation of the PGG, the impedance at the three ports shown in Figure 2.33 is calculated by Equation 2.105 (denoted by "semi-analytical method"). The results are compared with the full-wave simulation of the original PGG (denoted by "full-wave simulation") up to 100 GHz, as shown in Figure 2.36.

It can be seen that the self-impedance at port 1 can be well predicted by the analytical formula in Equation 2.105. However, there exists some deviation for impedances at port 2 and port 3. For impedance at port 2, we can see that the first-series resonant frequency obtained by the semi-analytical method is lower than that obtained by the full-wave simulation, whereas the parallel resonant frequencies of both methods agree well. For port 3, the difference of the first-series resonant frequency obtained by the two methods becomes even larger. The reason for this disagreement is that the inductance in the PGG changes to be larger when the

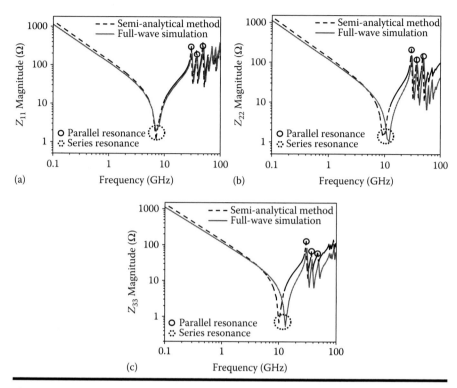

Figure 2.36 **Impedance obtained from the resonant cavity method (for the equivalent PGP) and the full-wave method (for the power–ground grid). Self-impedance at different ports: (a) port 1, (b) port 2, and (c) port 3.**

position of the port moves from the edge to the center [17]. To solve this problem, an empirical formula is proposed to modify the resonant cavity mode ϕ_{mni} and ϕ_{mnj} of the equivalent PGPs to ensure that it obtains the same impedance curve as that of the original PGG.

2.5.2 Modified Mode Function

According to the equivalent circuit in Figure 2.18, Z_{ji} can be represented by the equivalent circuit in Figure 2.37. When frequency is below the first parallel resonant frequency, the first-series resonant frequency can be approximated as the series of C_0 and summation of inductors in Figure 2.37.

$$f_s = \frac{1}{2\pi \sqrt{C_0 \displaystyle\sum_{m,n=0,m+n\neq0}^{\infty} L_{mn} N_{mni} N_{mnj}}} \quad (2.106)$$

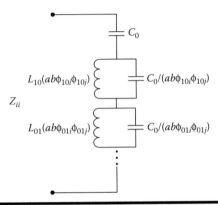

Figure 2.37 The equivalent circuit of the power–ground plane.

while the parallel resonant frequency is

$$f_{mn} = \frac{1}{2\pi\sqrt{L_{mn}C}} \tag{2.107}$$

As shown in Equations 2.106 and 2.107, the first-series resonant frequency is defined by ϕ_{mni}, ϕ_{mnj}, L_{mn}, and C_0, whereas the parallel resonant frequencies are defined by only L_{mn} and C_0. To compensate the difference of the first-series resonant frequency between equivalent PGP and the original PGG in Figure 2.36 and, at the same time, to keep the parallel resonant frequencies unchanged, an empirical formula is derived to modify the mode functions as follow:

$$z_{ji} \approx \sum_{m=0}^{\infty}\sum_{n=0}^{\infty} \frac{\sqrt{f_{mn}(x_i)f_{mn}(y_i)}\sqrt{f_{mn}(x_j)f_{mn}(y_j)}ab\phi_{mni}\phi_{mnj}}{(1/j\omega L_{mn}) + j\omega C_0}$$

$$= \sum_{m=0}^{\infty}\sum_{n=0}^{\infty} \frac{F_{mni}F_{mnj}N_{mni}N_{mnj}}{(1/j\omega L_{mn}) + j\omega C_0} \tag{2.108}$$

where

$$f_{mn}(x) = \begin{cases} 1 & m=0, \quad n=0 \\ -0.29\ln\left(\dfrac{2x}{a}+1\right)+1 & x \le \dfrac{a}{2}, \quad m\neq 0 \text{ or } n\neq 0 \\ -0.29\ln\left(\dfrac{2(a-x)}{a}+1\right)+1 & x > \dfrac{a}{2}, \quad m\neq 0 \text{ or } n\neq 0 \end{cases} \tag{2.109}$$

$$f_{mn}(y) = \begin{cases} 1 & m=0, \quad n=0 \\ -0.29\ln\left(\dfrac{2y}{b}+1\right)+1 & y\leq\dfrac{b}{2}, \quad m\neq0 \text{ or } n\neq0 \\ -0.29\ln\left(\dfrac{2(b-y)}{b}+1\right)+1 & y>\dfrac{b}{2}, \quad m\neq0 \text{ or } n\neq0 \end{cases} \quad (2.110)$$

and

$$F_{mni} = \sqrt{f_{mn}(x_i)f_{mn}(y_i)}, \quad F_{mnj} = \sqrt{f_{mn}(x_j)f_{mn}(y_j)} \quad (2.111)$$

$f_{mn}(x)$ and $f_{mn}(y)$ are the proposed empirical modified factors.

The first-series resonant frequency is modified by the empirical formula, which has no influence on the parallel resonant frequency. It will compensate the major difference at the series resonant frequency, as shown in Figure 2.36b and c. The accuracy of the proposed modified resonant cavity mode is validated in the next subsection.

2.5.3 Validation

In this subsection, the accuracy of the semi-analytical method with the modified resonant mode is validated with the full-wave method and an available measurement result. In the following figures, semi-analytical method with modified cavity mode is denoted as "modified semi-analytical method" for simplicity. The impedance of the three representatively positioned ports in Figure 2.33 is calculated. The coordinates of the ports are (0.05 mm, 0.05 mm) for port 1, (0.35 mm, 0.35 mm) for port 2, and (0.55 mm, 0.55 mm) for port 3. The simulated Z parameter for the frequency band of 0.1 to 100 GHz is plotted in Figure 2.38.

As shown in Figure 2.38, the modified resonant cavity mode for the equivalent planes illustrates the consistency with that of the full-wave method for PGG in the frequency range from 0.1 to 100 GHz. Meanwhile, the computing time with the modified semi-analytical method is much less than that with the full-wave method (several seconds vs. several hours). Therefore, the modified semi-analytical method can be used to calculate the impedance of the PGG efficiently and accurately.

Next, to validate the scalability of the proposed modified semi-analytical method, Z parameters of PGGs with different grid sizes are simulated. The grid patterns are with the size of L_w/L_p = 30 um/100 um and L_w/L_p = 40 um/100 um. The process of calculating the parameters of the equivalent PGP for the PGG is the same as depicted in the previous subsection. The equivalent medium of the

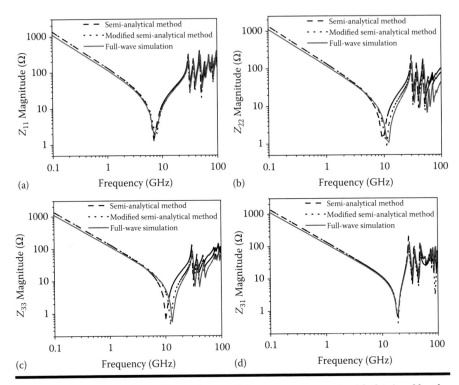

Figure 2.38 Comparison of impedance of the power–ground grid obtained by the semi-analytical method, modified semi-analytical method, and full-wave method. (a) Self-impedance at port 1, (b) self-impedance at port 2, (c) self-impedance at port 3, and (d) transfer impedance between port 1 and port 3.

equivalent PGP changes with the grid size, but the proposed empirical formula remains the same. Here, self-impedance of port 2 and port 3 are calculated for the validation. The results are shown in Figure 2.39.

As depicted in Figure 2.39, the result of the modified semi-analytical method agrees well with that of the full-wave method. It indicates that the proposed empirical formula is almost independent of the grid size. Therefore, the modified semi-analytical method can be a general method when calculating the PGG of typical sizes on interposer.

The electric field distribution of the modified (2,0) mode, $F_{20i}\phi_{20i}$, of the PGG is plotted in Figure 2.40. This mode distribution, together with the proposed

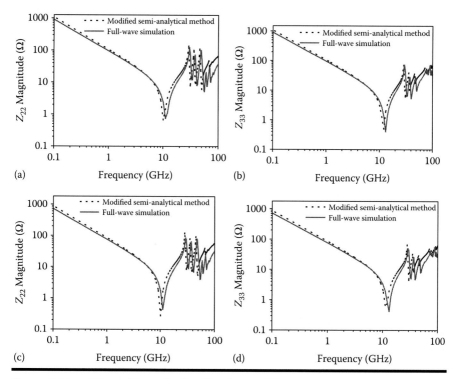

Figure 2.39   Z22 and Z33 obtained by the modified semi-analytical method and full-wave method for different power–ground grid sizes. (a) and (b) for *Lw/Lp* = 30 um/100 um, (c) and (d) for *Lw/Lp* = 40 um/100 um.

semi-analytical formula, is useful for the placement of the decoupling capacitors. For example, to eliminate the resonant frequency of the (2,0) mode, the decoupling capacitor should be placed in the region of maximum electric field.

Finally, the measurement of a practical on-chip PGG [18] is applied to validate the results obtained from the proposed modified semi-analytical method. The dimensions of the on-chip PDN can be found in [18]. In this example, the metal loss and substrate loss are considered in the proposed method by using a complex wavenumber k. As shown in Figure 2.41, there exists some deviation between the modified semi-analytical method and the measurement results. The deviation could be the result of the dimension variation, which is caused by the manufacturing process.

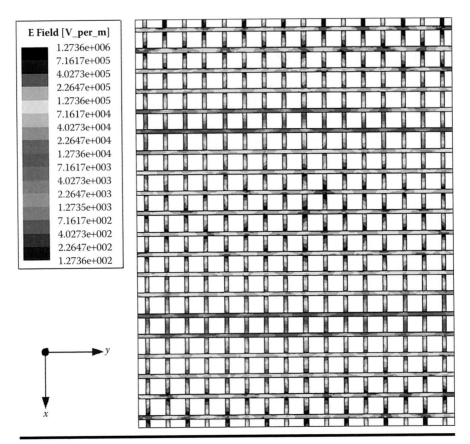

Figure 2.40 **Electric field distribution of the power–ground grid at resonant frequency of (2,0) mode.**

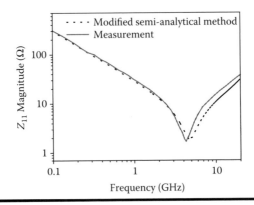

Figure 2.41 **Comparison of the impedance curves obtained by the modified semi-analytical method and the measurement. (From Pak, J. S., et al., *IEEE Trans. Compon. Pack. Manuf. Technol.*, 1(2), 208–219, 2011.)**

References

1. J. Kim, M. D. Rotaru, S. Baek, J. Park, M. K. Iyer, and J. Kim, Analysis of noise coupling from a power distribution network to signal traces in high-speed multilayer printed circuit boards, *IEEE Trans. Electromagn. Compat.*, 48(2), 319–330, 2006.
2. M. Swaminathan, J. Kim, I. Novak, and J. Libous, Power distribution networks for system-on-package: Status and challenges, *IEEE Trans. Adv. Packag.*, 27(2), 286–300, 2004.
3. X. C. Wei, E. P. Li, E. X. Liu, and R. Vahldieck, Efficient simulation of power distribution network by using integral equation and modal decoupling technology, *IEEE Trans. Microw. Theory Technol.*, 56(10), 2277–2285, 2008.
4. A. E. Engin, W. John, G. Sommer, W. Mathis, and H. Reichl, Modeling of striplines between a power and a ground plane, *IEEE Trans. Adv. Pack.*, 29(3), 415–426, 2006.
5. SIWave, ANSYS software, http://www.ansys.com.
6. L. Tsang, H. Chen, C. C. Huang, and V. Jandhyala, Modeling of multiple scattering among vias in planar waveguides using Foldy-Lax equations, *Micro. Opt. Technol. Lett.*, 31(3), 201–208, 2001.
7. D. C. Yang, X. C. Wei, X. C. Zhang, L. S. Zhang, and E. P. Li, A novel hybrid analytical method for impedance calculation of power and ground planes, *IEEE Trans. Electromagn. Compat.*, 55(5), 949–955, 2013.
8. D. C. Yang and X. C. Wei, Impedance calculation of power and ground planes by using imaging methods, *Asia-Pacific Symposium on Electromagnetic Compatibility*, Singapore, May 21–24, 2012.
9. H. Q. Ye, X. C. Wei, and E. P. Li, A novel semi-analytical solution of impedance of grid-type power distribution network, *IEEE International Symposium on Electromagnetic Compatibility and EMC Europe*, Dresden, August 16–22, 2015.
10. T. L. Wu, Y. H. Lin, T. K. Wang, C. C. Wang, and S. T. Chen, Electromagnetic band-gap power/ground planes for wideband suppression of ground bounce noise and radiated emission in high-speed circuits, *IEEE Trans. Microw. Theory and Technol.*, 53(9), 2935–2942, 2005.
11. R. L. Chen, J. Chen, and T. H. Hubing, Analytical model for the rectangular power-ground structure including radiation loss, *IEEE Trans. Electromagn. Compat.*, 47(1), 10–16, 2005.
12. U. Carlberg, P. S. Kildal, and J. Carlsson, Study of antennas in reverberation chamber using method of moments with cavity Green's function calculated by Ewald summation, *IEEE Trans. Electromagn. Compat.*, 47(4), 805–814, 2005.
13. M. J. Park, Accelerated summation of the Green's function for the rectangular cavity, *IEEE Micro. Comp. Lett.*, 19(5), 260–262, 2005.
14. M. Abramowitz and I. A. Stegun, *Handbook of Mathematical Functions with Formulas, Graphs, and Mathematical Tables.* Chapter 6, Dover Publications, 1972.
15. F. Capolino, D. R. Wilton, and W. A. Johnson, Efficient computation of the 2-D Green's function for 1-D periodic structures using the Ewald method, *IEEE Trans. Anten. Propag.*, 53(9), 2977–2984, 2005.
16. K. Kim, J. M. Yook, J. Kim, H. Kim, J. Lee, K. Park, and J. Kim, Interposer power distribution network (PDN) modeling using a segmentation method for 3-D ICs with TSVs, *IEEE Trans. Compon. Pack. Manuf. Technol.*, 3(11), 1891–1906, 2013.

17. J. S. Pak, J. Kim, J. Cho, J. Lee, H. Lee, K. Park, and J. Kim, On-chip PDN design effects on 3D stacked on-chip PDN impedance based on TSV interconnection, *IEEE Electrical Design of Advanced Packaging & Systems Symposium*, Singapore, December 7–9, 2010.
18. J. S. Pak, J. Kim, J. Cho, K. Kim, T. Song, S. Ahn, J. Lee, H. Lee, K. Park, and J. Kim, PDN impedance modeling and analysis of 3D TSV IC by using proposed P/G TSV array model based on separated P/G TSV and chip-PDN models, *IEEE Trans. Compon. Pack. Manuf. Technol.*, 1(2), 208–219, 2011.

Chapter 3

Integral Equation Solutions

This chapter introduces the integral equation solutions of the power distribution network (PDN). The integral equation method, finite element method, and finite difference method are the three major electromagnetic algorithms for all kinds of electromagnetic modeling. For the PDN, finite element method and finite difference method had been well studied, and there had been related commercial software, such as the SIWave (from ANSYS company) and Sigrity (from Cadence company), and the algorithm discussed in Section 1.2.3. However, there are few applications of integral equation methods on the PDN, and recently new works have been published on this topic.

The key of integral equation methods is the Green's function of the structure under study. The content of this chapter can be taken as the extension of the Green's function of Chapter 1 to arbitrarily shaped power–ground planes. The integral equation methods are classified into two kinds according to their different Green's functions as shown in Figure 3.1: the two-dimensional (2D) integral equation solution of the PDN, where the electromagnetic field is assumed uniform along the vertical direction, and the three-dimensional (3D) integral equation solution, where the electromagnetic field forms standing wave along the vertical direction.

Because the dielectric constant of the substrate is larger than that of the surrounding air, the boundary of the power–ground planes can be taken as a perfect magnetic conductor (PMC), and the power–ground planes form a cavity. When the power–ground planes have regular shapes (such as rectangular and triangular), for 2D/3D integral equation solution, an analytic Green's function can be obtained for such cavity. The 2D solution for rectangular power–ground planes had been discussed in Chapter 1.

There is no analytic Green's function for arbitrarily shaped power–ground planes. To solve this problem, the equivalence principle is used to extend the finite-size

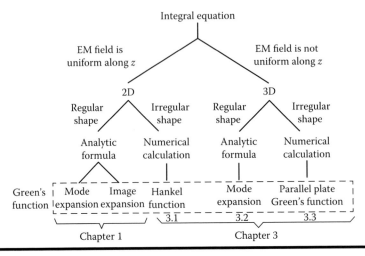

Figure 3.1   Integral equation solutions of the power–ground planes.

power–ground planes to infinite parallel plates, so that the Green's function of infinite parallel plates can be used. The discontinuities in the original power–ground planes, including the slot, antipad, and boundary, are represented by equivalent electric and magnetic currents. Those equivalent electric and magnetic currents are unknowns and are numerically solved. In 2D solution, the Hankel function is used as the Green's function. In 3D solution, the dyadic Green's functions are employed. The solution of these integral equations results in an equivalent microwave network of the power–ground plane pair. The application of this equivalent network together with other equivalent circuits for the system-level analysis is presented. By comparing the integral equation methods with measurement results and available full-wave methods, their accuracy and efficiency are verified.

Finally, a combined equivalent networks analysis for the power–ground planes with narrow slots is presented. Here the whole structure is decoupled into two subdomains: the power and ground planes and the slots, which support the parallel-plate mode and the slot mode, respectively. 3D integral equations are created on each subdomain and equivalent networks are extracted through the moment method solution of the created integral equations. These two equivalent networks are connected together through the ports defined on the interface between two subdomains to consider the mode conversion between the parallel-plate mode and the slot mode.

3.1 2D Integral Equation Solution

To provide the low-impedance path and reduce the interference between circuits, the power–ground planes are widely employed in the PDN of the electronic package and printed circuit board. However, the power–ground planes also introduce some electromagnetic compatibility problems due to the ever-increasing operating frequency and power density as discussed in Chapter 1. Efficient and accurate electrical modeling technologies are highly demanded to tackle this challenge. The cavity mode theory [1] discussed in Chapter 1 gives a quick analytic simulation of the power–ground planes. However, it is limited to the power–ground planes with regular shapes, such as the rectangle and triangle. Other semi-analytical methods, such as the cylindrical wave expansion method [2,3], also has the difficulty in modeling the power–ground planes with arbitrary shapes. More accurate modeling of the power–ground planes requires directly solving the electromagnetic field by using 3D numerical methods, such as the partial element equivalent circuit (PEEC) method [4] and the finite differential method [5,6]. Although the overall electrical size of the power–ground planes is small enough to apply the 3D methods, the high respect ratio of the power–ground planes results in a very dense meshing. This makes these 3D methods very expensive in terms of computing time and memory requirements.

The hybrid methods [7,8] show the ability of accurate and efficient simulations of electromagnetic problems, because they benefit from both analytic and numerical techniques. It had been demonstrated that the electromagnetic field between a pair of conducting planes can be decomposed and treated separately [9]. In the work of Wei et al. [10], a hybrid method is proposed to simulate the emission and susceptibility of the power–ground planes with rectangular shapes. In this section, a simple yet accurate modal decoupling method is proposed to simulate the power–ground planes with arbitrary shapes. The total electromagnetic field of the PDN is decoupled into the parallel-plate mode and the transmission line mode. The transmission line mode includes the stripline mode and the microstrip line mode. The parallel-plate mode is solved by using an efficient 2D integral equation method. Meanwhile, the discontinuity due to the through-hole via is also considered by an analytic formula.

Finally, the whole equivalent circuit of the PDN is obtained, which can be substituted into a circuit simulator to perform the system-level electromagnetic compatibility analysis of the whole PDN.

This proposed method is named as the integral equation equivalent circuit (IEEC) method. The advantages of the proposed 2D method over available 3D methods are as follows:

1. It decouples the complex 3D problem into two simple 2D problems. Therefore, it greatly reduces the computing time and still keeps a good accuracy.
2. It extracts the equivalent circuit from the complex electromagnetic field distribution. This equivalent circuit provides a more comprehensive solution for the real industrial applications than the purely electromagnetic field solvers.

3.1.1 Formulation

Figure 3.2 shows the typical structure of the PDN. The antipad is a clearance hole between the via and the metal planes. The ground and power planes are highly conducting metal planes, which provide a low-impedance path for the power supply.

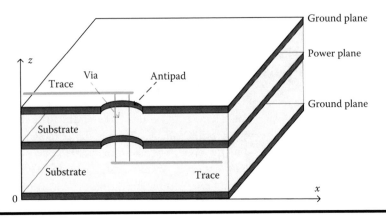

Figure 3.2   The typical structure of the PDN.

Usually, the substrate sandwiched between metal planes is uniform, isotropic, and with a thickness much smaller than the interesting wavelength. Based on this, we decouple the total electromagnetic field into two kinds of independent modes: the parallel-plate mode and the transmission line mode.

1. The parallel-plate mode represents the standing-wave field constrained between the ground and power planes. It is due to the reflections from the edges of the cavity-like power–ground plane pair. Due to the thin substrate, it is assumed that the electromagnetic field does not change along the z-direction. The vector electric field E is in the z-direction, and the vector magnetic field H lies in the xoy plane.
2. The signal traces sandwiched between the power–ground planes are taken as the striplines, whereas the signal traces above or below the power–ground planes are taken as the microstrip lines. They support the transmission line modes. They are transverse electromagnetic (TEM) modes (transverse to the traces' directions) and propagate along the traces.

These two kinds of modes can be solved in the xoy plane and the cross section of traces separately. After that, the parallel-plate mode converses with the transmission line mode at the through-hole via region. How to calculate the through-hole via's equivalent circuit is a very interesting topic of a long history [11–13]. In this section, a simple but accurate analytic formula is derived based on the work of Zhang et al. [14], to calculate the parasitic capacitance of the through-hole via.

According to the above modal decoupling, the whole PDN is decomposed into three subdomains as in Figure 3.3: the power–ground plane pair, the microstrip lines and striplines, and the through-hole vias. Their electromagnetic field distributions are also plotted in Figure 3.3.

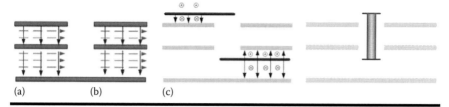

Figure 3.3 Three subdomains used for the modal decoupling: (a) power–ground plane pair, (b) signal traces, and (c) through-hole via. →: electric field; —>: magnetic field.

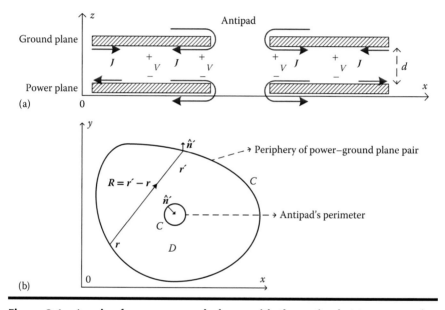

Figure 3.4 A pair of power–ground planes with the antipad: (a) cross section and (b) top view.

The transmission line modes can be easily solved by using available transmission line solvers. In the following, we will focus on the 2D integral equation solution of the power–ground planes, the recombination of the parallel-plate mode and the transmission line mode, and the whole equivalent circuit of the PDN including the through-hole via's effect.

3.1.1.1 Integral Equation Solution of the Power–Ground Planes

Figure 3.4a shows a pair of power–ground planes with the antipad. Because the electromagnetic field is independent with z, the power–ground planes can be modeled as a 2D problem. A 2D region D is defined in Figure 3.4b, where the periphery of the power–ground planes and all antipad's perimeters form the boundary C with

$\hat{n}'$ as its outward unit normal vector. According to the directions of the electromagnetic fields, we define the voltage between the top and bottom planes as $V = -dE_z$ with d being the thickness of the substrate, and the horizontal current density as $J = -\hat{z} \times H$ on the top plane and $J = \hat{z} \times H$ on the bottom plane. J on the top and bottom planes has the same values and opposite directions.

By using the integral equation theory [1], the voltage at a certain point of the contour C can be expressed as the integral of voltages and current densities along the whole contour C:

$$V(r) = \frac{k}{2j} \oint_C \left[\frac{R}{R} \cdot \hat{n}' H_1^{(2)}(kR) V(r') + j\eta d H_0^{(2)}(kR) J_n(r') \right] dl' \quad (3.1)$$

where $H_0^{(2)}$ and $H_1^{(2)}$ are the zero-order and first-order Hankel functions of the second kind, respectively. $R = r' - r$ and R represents the length of R. r and r' are the observation and source points, respectively, located on C. The prime on dl emphasizes that the integration is over r'. k and η are the wave number and the wave impedance of the substrate, respectively $k = \omega\sqrt{\mu\varepsilon}(1 - j(\tan\delta + t/d)/2)$ and $\eta = \omega\mu/k$, where $\tan\delta$ is the loss tangent of the substrate and t is the skin depth of the planes $j = \sqrt{-1}$ $J_n = \hat{n}' \cdot (\hat{z} \times H)$ means the current density flowing into/from the region D on the top/bottom plane.

(V, J_n) in Equation 3.1 is classified into (V_p, J_p) (along the periphery) and (V_a, J_a) (along the perimeters of the antipads). The physical meaning of J_a is explained in Figure 3.5. I_{trace} denotes the current along the signal trace. J_a starts from the top plane; passes through the distributed resistance, inductance, capacitance, and conductance between the top and bottom planes; and arrives at the bottom plane. Therefore, J_a represents the displacement return current density of I_{trace}.

The periphery of the power–ground planes is divided into many straight segments as those in [1]. The antipad's perimeter is much small in terms of the interesting wavelength; V_a along each antipad's perimeter is assumed to be constant. However, due to the current proximity effect at the high frequency, J_a is not uniformly distributed along the antipad's perimeter. To accurately model J_a, each antipad's perimeter is divided into four small segments. After that, (V, J_n) in

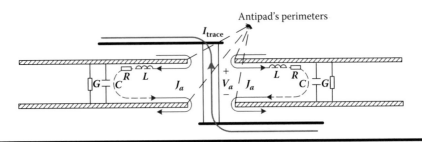

Figure 3.5 **The re-routed return current of the traces (side view).**

Equation 3.1 is expanded by using unit pulse functions defined on these segments and antipads' perimeters:

$$V_p(\mathbf{r}) = \sum_{i=1}^{N_p} V_{p,i} P_{p,i}(\mathbf{r}) \text{ and } V_a(\mathbf{r}) = \sum_{i=1}^{N_a} V_{a,i} P_{a,i}(\mathbf{r}) \tag{3.2}$$

$$J_p(\mathbf{r}) = \sum_{i=1}^{N_p} I_{p,i} P_{p,i}(\mathbf{r})/w_{p,i} \text{ and } J_a(\mathbf{r}) = \sum_{i=1}^{N_a} I_{a,i} P_{a,i}(\mathbf{r})/w_{a,i} \tag{3.3}$$

with $P_{p/a,i}(\mathbf{r}) = \begin{cases} 1, \mathbf{r} \in w_{p/a,i} \\ 0, \mathbf{r} \notin w_{p/a,i} \end{cases}$. $w_{p/a,i}$ represents the ith segment/antipad's perimeter.

In the following, $w_{p/a,i}$ is also used to represent the length of the ith segment/antipad's perimeter. $I_{p/a,i}$ is the current flowing on the ith segment/antipad's perimeter. $N_{p/a}$ is the number of peripheral segments/antipads.

After this expansion, we define ports on each peripheral segment and antipad's perimeter as in Figure 3.6. Let $[V_p V_a]$ and $[I_p I_a]$ represent the expansion coefficients in Equations 3.2 and 3.3, the whole power–ground plane pair is equivalent to an N-port circuit network with $[V_p V_a]$ and $[I_p I_a]$ as its port voltages and currents, respectively. $N = N_a + N_p$. It should be noted that the reference planes of these ports are defined on the internal surfaces of the power and ground planes. Because each antipad's perimeter is divided into four small segments, it is equivalent to four ports. Because V_a along each antipad's perimeter is constant, these four ports are parallel connected and reduced to one port.

Substituting Equations 3.2 and 3.3 into 3.1, and matching both sides of Equation 3.1 with $\mathbf{r}$ at the center of each segment, the following N-by-N linear equations are obtained:

$$[U] \cdot \begin{bmatrix} V_p \\ V_a \end{bmatrix} = [H] \cdot \begin{bmatrix} I_p \\ I_a \end{bmatrix} \tag{3.4}$$

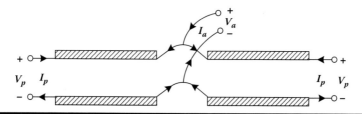

Figure 3.6 Definitions of port voltages and currents (cross section).

where:

$$[U] = \begin{bmatrix} U^{pp} & U^{pa} \\ U^{ap} & U^{aa} \end{bmatrix} \quad \text{and} \quad [H] = \begin{bmatrix} H^{pp} & H^{pa} \\ H^{ap} & H^{aa} \end{bmatrix}$$

The elements in $[U]$ and $[H]$ are calculated as follows:

$$U_{ij} = \delta_{ij} - \frac{k}{2j} \int_{w_{pla,j}} \frac{R}{R} \cdot \hat{n}'_j H_1^{(2)}(kR) dl'_j \tag{3.5}$$

$$H_{ij} = \frac{k\eta d}{2w_{pla,j}} \int_{w_{pla,j}} H_0^{(2)}(kR) dl'_j \tag{3.6}$$

where:
    $\hat{n}'_j$ is the normal vector of the jth peripheral segment/antipad
    δ_{ij} is the Kronecker delta

For $I = j$ or R is very small, the Hankel functions in Equations 3.5 and 3.6 are singular or near singular. In this case, to give a more accurate result, the above integrals are analytically calculated in the appendix.

A perfect magnetic wall is assumed along the periphery due to the thin substrate. It means $[I_p] = 0$ in Equation 3.4, so we can get

$$[Z^a] \times [I_a] = [V_a] \tag{3.7}$$

$$[Z^a] = \left([U^{aa}] - [U^{ap}] \cdot [U^{pp}]^{-1} \cdot [U^{pa}] \right)^{-1} \left([H^{aa}] - [U^{ap}] \cdot [U^{pp}]^{-1} \cdot [H^{pa}] \right) \tag{3.8}$$

The elements in $[Z^a]$ represent the self and mutual ground impedances of the power–ground planes. In this case, each power–ground planes pair is equivalent to a network with $[Z^a]$ as its impedance matrix. It should be noted that this proposed method can solve the power–ground planes with arbitrary shapes because C in Equation 3.1 can be arbitrarily shaped.

3.1.1.2 Recombination of Stripline–Parallel-Plate Mode

In this section, the above-obtained equivalent network of power–ground planes is connected with the equivalent network of the striplines. For the sake of simplicity, a single stripline sandwiched between a pair of power–ground planes as shown in Figure 3.7 is used here to demonstrate the recombination of the parallel-plate mode and the stripline mode. The proposed recombination method can be easily extended to multi-striplines cases.

For the stripline commonly used in microwave engineering, its two parallel reference planes are shorted and equipotential. For the stripline sandwiched between

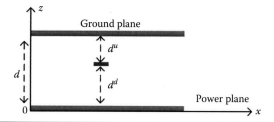

Figure 3.7　Cross section of a stripline.

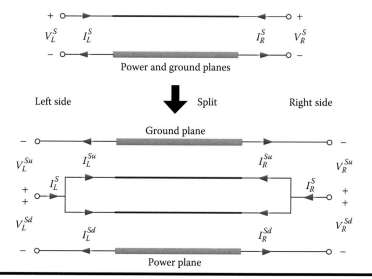

Figure 3.8　One stripline is split into up and down striplines.

the power and ground planes, the parallel-plate modes are excited which accumulate voltage drop between the power and ground planes. To consider the potential difference between the power and ground planes, we split the stripline and accordingly divide the total current into the currents flowing on the up and down surfaces of the stripline separately. Figure 3.8 shows such splitting. The subscripts L and R denote the left and right ports of the transmission lines, respectively. The superscript S denotes the unsplit stripline; S_u and S_d denote the split up and down striplines, respectively.

For the unsplit stripline, the admittance matrix is defined as

$$\begin{bmatrix} I_L^S \\ I_R^S \end{bmatrix} = \begin{bmatrix} \boldsymbol{Y}^S \end{bmatrix} \begin{bmatrix} V_L^S \\ V_R^S \end{bmatrix} \tag{3.9}$$

For the split stripline, the up and down admittance matrices are defined as

$$\begin{bmatrix} I_L^{S_u} \\ I_R^{S_u} \end{bmatrix} = \begin{bmatrix} \mathbf{Y}^{S_u} \end{bmatrix} \begin{bmatrix} V_L^{S_u} \\ V_R^{S_u} \end{bmatrix} \tag{3.10}$$

$$\begin{bmatrix} I_L^{S_d} \\ I_R^{S_d} \end{bmatrix} = \begin{bmatrix} \mathbf{Y}^{S_d} \end{bmatrix} \begin{bmatrix} V_L^{S_d} \\ V_R^{S_d} \end{bmatrix} \tag{3.11}$$

where $V_{R/L}^{S/S_u/S_d}$ and $I_{R/L}^{S/S_u/S_d}$ are port voltages and currents defined in Figure 3.8. $I_{R/L}^{S_u}$ and $I_{R/L}^{S_d}$ flow on the up and down surfaces of the stripline respectively.

The split stripline returns to the unsplit stripline when the power and ground planes are shorted. By shorting the power and ground planes of the split stripline, we get

$$\begin{bmatrix} V_L^{S_u} \\ V_R^{S_u} \end{bmatrix} = \begin{bmatrix} V_L^{S_d} \\ V_R^{S_d} \end{bmatrix} = \begin{bmatrix} V_L^{S} \\ V_R^{S} \end{bmatrix} \tag{3.12}$$

When the stripline approaches the ground plane ($d^u \to 0$ in Figure 3.7), $I_{R/L}^{S_u} \to I_{R/L}^{S}$ and $I_{R/L}^{S_d} \to 0$. Conversely, when the stripline approaches the power plane ($d^u \to d$ in Figure 3.7), $I_{R/L}^{S_d} \to I_{R/L}^{S}$ and $I_{R/L}^{S_u} \to 0$. Based on this observation and that d is much smaller, we assume that

$$\begin{bmatrix} I_L^{S_u} \\ I_R^{S_u} \end{bmatrix} = \frac{d^d}{d} \begin{bmatrix} I_L^{S} \\ I_R^{S} \end{bmatrix} \quad \text{and} \quad \begin{bmatrix} I_L^{S_d} \\ I_R^{S_d} \end{bmatrix} = \frac{d^u}{d} \begin{bmatrix} I_L^{S} \\ I_R^{S} \end{bmatrix} \tag{3.13}$$

Substituting Equations 3.12 and 3.13 into 3.9 through 3.11, we get

$$[\mathbf{Y}^{S_u}] = \frac{d^d}{d} [\mathbf{Y}^{S}] \tag{3.14}$$

$$[\mathbf{Y}^{S_d}] = \frac{d^u}{d} [\mathbf{Y}^{S}] \tag{3.15}$$

Observe that $[\mathbf{Y}^{S}] = [\mathbf{Y}^{S_u}] + [\mathbf{Y}^{S_d}]$.

For the structure shown in Figure 3.7, it can be equivalent to the combination of three networks: the up split stripline, the down split stripline, and the equivalent network of power–ground planes. Figure 3.9 shows the port voltages and currents of these three networks. Figure 3.10 shows their connections. The combined network is a four-port network with the admittance matrix defined as

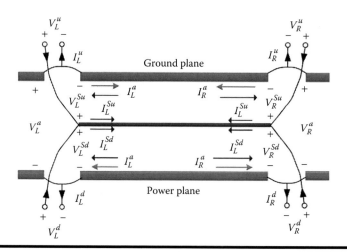

Figure 3.9 **Port voltages and currents of three equivalent networks.**

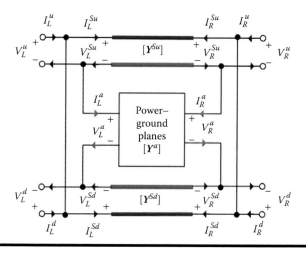

Figure 3.10 **Recombination of the parallel-plate mode and the stripline mode.**

$$
\begin{bmatrix} I_L^u \\ I_R^u \\ I_L^d \\ I_R^d \end{bmatrix} = [Y] \begin{bmatrix} V_L^u \\ V_R^u \\ V_L^d \\ V_R^d \end{bmatrix} \tag{3.16}
$$

with port voltages $V_{R/L}^{u/d}$ and currents $I_{R/L}^{u/d}$ defined in Figures 3.9 and 3.10. These four ports account for the possible connections between this stripline and external microstrip lines or striplines.

For the power–ground planes, the admittance matrix is defined as

$$
\begin{bmatrix} I_L^a \\ I_R^a \end{bmatrix} = \begin{bmatrix} Y^a \end{bmatrix} \begin{bmatrix} V_L^a \\ V_R^a \end{bmatrix}
\tag{3.17}
$$

with $[Y^a] = [Z^a]^{-1}$.

From Figures 3.9 and 3.10, we can get

$$
\begin{bmatrix} \begin{bmatrix} V_L^{S_u} \\ V_R^{S_u} \end{bmatrix} \\ \begin{bmatrix} V_L^{S_d} \\ V_R^{S_d} \end{bmatrix} \\ \begin{bmatrix} V_L^a \\ V_R^a \end{bmatrix} \end{bmatrix} = [T] \cdot \begin{bmatrix} \begin{bmatrix} V_L^u \\ V_R^u \end{bmatrix} \\ \begin{bmatrix} V_L^d \\ V_R^d \end{bmatrix} \end{bmatrix} \quad \text{and} \quad \begin{bmatrix} \begin{bmatrix} I_L^u \\ I_R^u \end{bmatrix} \\ \begin{bmatrix} I_L^d \\ I_R^d \end{bmatrix} \\ \begin{bmatrix} I_L^a \\ I_R^a \end{bmatrix} \end{bmatrix} = [T]^t \cdot \begin{bmatrix} \begin{bmatrix} I_L^{S_u} \\ I_R^{S_u} \end{bmatrix} \\ \begin{bmatrix} I_L^{S_d} \\ I_R^{S_d} \end{bmatrix} \\ \begin{bmatrix} I_L^a \\ I_R^a \end{bmatrix} \end{bmatrix}
\tag{3.18}
$$

with $[T] = \begin{bmatrix} U & 0 \\ 0 & U \\ -U & U \end{bmatrix}$, and $[U]$ is a 2×2 unit matrix. Superscript t in the above

equation means the transpose.

Substituting Equations 3.10, 3.11, 3.17, and 3.18 into 3.16, we get the admittance matrix of the recombined network as

$$
[Y] = [T]^t \cdot \begin{bmatrix} Y^{S_u} & 0 & 0 \\ 0 & Y^{S_d} & 0 \\ 0 & 0 & Y^a \end{bmatrix} \cdot [T] = \begin{bmatrix} Y^{S_u} + Y^a & -Y^a \\ -Y^a & Y^{S_d} + Y^a \end{bmatrix}
\tag{3.19}
$$

It should be mentioned that the proposed recombination method can be easily extended to include multi-striplines. In that case, scalars $V_{R/L}^{S/S_u/S_d/u/d/a}$ and $I_{R/L}^{S/S_u/S_d/u/d/a}$ will be extended to vectors.

3.1.1.3 Equivalent Circuit of Through-Hole Via

The through-hole via is an important discontinuous structure in the PDN. Figure 3.11 shows a through-hole via and it equivalent power integrity (PI) circuit.

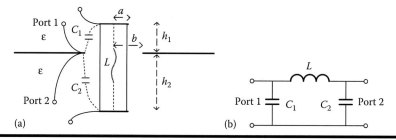

Figure 3.11 (a) Through-hole via (cross section) and (b) its equivalent PI circuit.

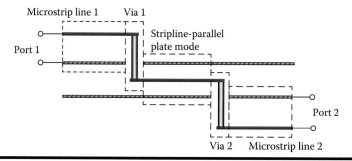

Figure 3.12 A PDN.

The closed form of L in Figure 3.12 can be obtained from [13]. Measurements had indicated that the effect of through-hole via is mainly capacitive, so the parasitic capacitors C_1 and C_2 should be accurately calculated. By assuming $R \to \infty$ in [14] and using the asymptotic expansions of Bessel functions, we get

$$C_{1/2} = \frac{4\pi\varepsilon}{\left[h_{1/2}\ln\left(b/a\right)\right]} \sum_{n=1,3,5,\ldots}^{2M-1} \frac{\left[1 - K_0\left(k_n b\right)/K_0\left(k_n a\right)\right]}{k_n^2} \tag{3.20}$$

Where K_0 are the modified Bessel function of the second kind with zero order. $k_n = \sqrt{(n\pi/2h_{1/2})^2 - 1/\lambda_g^2}$ with λ_g being the wavelength in the substrate.

3.1.1.4 Recombination of the Microstrip Line–Parallel-Plate Mode and the Whole Equivalent Circuits

The microstrip line mode can be easily combined with the parallel-plate mode through the through-hole via's equivalent circuit shown in Figure 3.11b. Finally, the whole equivalent circuits of the PDN can be obtained. Figure 3.12 shows a typical PDN. To get its equivalent circuit,

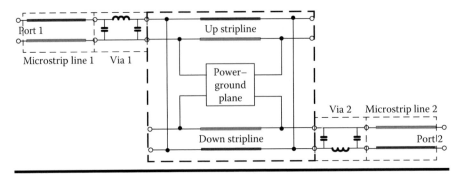

Figure 3.13   Equivalent circuit of the PDN of Figure 3.12.

1. Divide the whole structure into five parts: microstrip line 1, via 1, stripline–parallel-plate mode, via 2, and microstrip line 2.
2. Each part is equivalent to a network with its ports defined in Figures 3.9 and 3.11.
3. Calculate the *ABCD* matrices of each network and then cascade them to get the final equivalent circuit as in Figure 3.13.

3.1.2 Validations and Discussions

3.1.2.1 Ground Impedance of a Power–Ground Planes Pair

In this example, the ground impedance matrix $[Z^a]$ obtained from Equation 3.8 is validated. The power–ground planes under study are shown in Figure 3.14. The substrate between the metal planes has a thickness of 1.2 mm, a dielectric constant of 4.1, and a loss tangent of 0.015. Three subminiature A (SMA) connectors are mounted on the up plane. The inner conductors of the SMAs pass through the antipads on the up plane and are soldered to the down plane. The outer conductors of the SMAs are soldered to the up plane. The network analyzer is then connected to these SMAs to measure the *S* parameters. Figure 3.15 shows S_{11} and S_{21} obtained

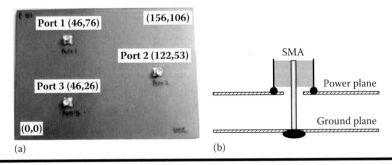

Figure 3.14   A pair of power–ground planes with three SMAs mounted on it (unit: mm): (a) top view and (b) cross section.

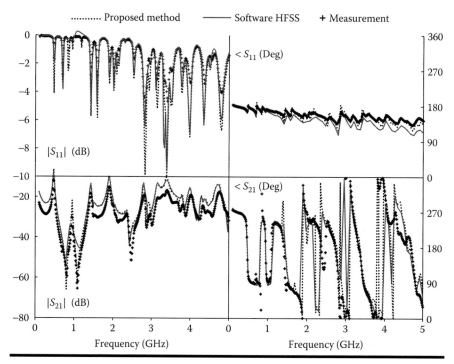

Figure 3.15 **S-parameters of the power–ground planes.** *Left*: **magnitude;** *right*: **phase.**

by the proposed method, measurement, and the 3D numerical software HFSS (from ANSYS company). Good agreement can be observed from the figures. For the proposed method, the $[Z^a]$ is converted to S parameters by using the reference impedance of 50 Ω. During the simulation, the power and ground planes are modeled as perfect conductor sheets with zero thickness, because their thickness is much smaller than the wavelength.

3.1.2.2 S Parameters of a Signal Trace

The power–ground plane pair from [11] is shown in Figure 3.16 for the validation. The signal trace includes three parts: left microstrip line, stripline, and right microstrip line. Two through-hole vias are used to connect them. Three ports are defined in Figure 3.16: port 1 is located at an upper position and connected to the up plane and the down plane, ports 2 and 3 are connected to the two ends of the trace, respectively. Port 1 is used to generate the noise between the two planes, and the coupling between port 1 and port 2 is analyzed. In the simulation, the thickness of the trace and planes is 0.01 mm and the dielectric constant of the substrate is 4.6.

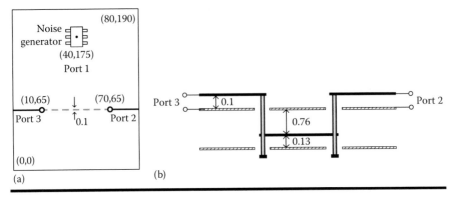

Figure 3.16 **Dimensions of the signal trace passing through the power–ground planes (unit: mm): (a) top view and (b) side view.**

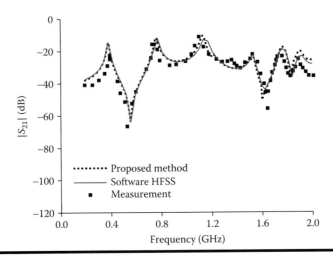

Figure 3.17 **Coupling coefficient between port 1 and port 2.**

Figure 3.17 shows the magnitude of S_{21} obtained by the proposed method, the measurement result from [11], and the software HFSS. Good agreement can be observed again. The peaks of S_{21} are due to the strong resonances of the power–ground planes. At these resonant frequencies, the noise coupling between port 1 and port 2 is pronounced. In the practical design, the decoupling capacitors are usually used to eliminate the resonances of the power–ground planes.

3.1.2.3 S Parameters of Two Coupled Signal Traces

The final example is two coupled signal traces. Their dimensions are shown in Figure 3.18. The substrate has a dielectric constant of 4.1. The reflection, transmission, and cross talk characteristics of the traces are simulated in Figure 3.19.

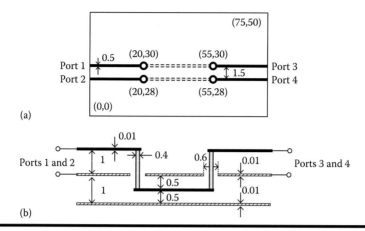

(a)

(b)

Figure 3.18 **Two coupled signal traces passing through the power–ground planes (unit: mm): (a) top view and (b) side view. The thickness of all traces and planes is 0.01 mm.**

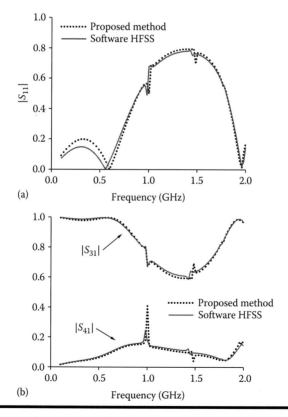

(a)

(b)

Figure 3.19 **(a) Reflection and (b) transmission and cross talk characteristics of the coupled signal traces.**

Table 3.1   Computing Times for the Proposed Method and 3D Simulator HFSS

Example	Proposed Method	HFSS
3.1.2.1	39.5 S	5 minutes
3.1.2.2	41 S	70 minutes
3.1.2.3	15 S	20 minutes

Good agreement for the cross talk $|S_{41}|$ is observed. This demonstrates the accuracy of the proposed method for the simulation of closely placed multi-traces. Both the transmission line mode and the parallel-plate mode contribute to the cross talk. And at the power–ground planes' inherent resonant frequency, this cross talk is amplified.

3.1.2.4 Computing Time Comparison

Table 3.1 lists the computing times of the proposed method and the 3D simulator HFSS. The computing time of the proposed method includes the solution of the parallel-plate mode and the transmission line mode, and the combination of their equivalent circuits. It can be seen that the proposed method greatly reduces the computing time compared with the HFSS. This is because that the most time-consuming part of the proposed method is the numerical solution of the integral equation along the one-dimensional contour C. Therefore, the number of unknowns is dramatically reduced. However, for the HFSS, a dense volume meshing is required because of the large aspect ratio of the power–ground planes and the tiny structures such as the through-hole and traces. This results in a lot of unknowns.

3.1.3 Conclusion

A 2D IEEC method is proposed in this section for the simulation of high-speed PDN. It decomposes the complex 3D problem into two simple 2D problems: the parallel-plate mode problem and the transmission line mode problem. Therefore, it greatly reduces the computing time and still keeps a good accuracy. The equivalent circuit is extracted from the modal analysis, which can be substituted into a circuit simulator to perform the signal integrity and power integrity cosimulations.

3.2 3D Integral Equation Solution

The simultaneous switching noise (SSN) is one of the critical electromagnetic compatibility (EMC) issues for the high-speed circuit. It is responsible for the cross talk between different signal traces and unwanted emission. Such emission leaks into the surrounding area of the package through the periphery and gaps of the power–ground planes. The electromagnetic susceptibility is another critical EMC problem, which may lead to the malfunction of the circuit integrated inside the package. Above electromagnetic emission and susceptibility problems are even more pronounced, if the noise spectrum covers any of the resonance frequencies of the power–ground planes.

These challenges require the simultaneous simulations of emission and susceptibility. However, established electromagnetic solvers face great difficulties due to accuracy or memory limitations. The cavity mode theory [1,15] provides an efficient analytic solution of the power–ground planes' impedance, provided that the periphery can be assumed to be a PMC and the electromagnetic fields are uniform along the vertical direction. More general modeling of the power–ground planes' structure at high frequencies can be achieved with full-wave numerical methods. The most popular techniques include the PEEC method [4] and the finite-difference time-domain method. However, these techniques are, in general, expensive in terms of central processing unit (CPU) and memory requirements.

To alleviate the computational cost of full-wave methods, algorithms that exploit analytic representations of features of the power–ground planes have been proposed. Among them, the Foldy-Lax multiple scattering method appears to be the most popular one. In [2], this method was used to analyze the vias in ball grid arrays, and its superior performances with respect to a universal full-wave solver (HFSS [16]) had been demonstrated. In [17], this method had been extended to simulate the power–ground planes with PMC boundaries of circular shape. In [18], this method was coupled with the layered dyadic Green's function in order to consider the external area of the power–ground planes.

These studies have made the assumptions that the power–ground planes are continuous and extend to infinity [1,18] or that their boundaries are circular PMCs [17]. However, in reality, the most common power–ground planes are finite with a rectangular boundary. If the power–ground planes are assumed to be infinitely extended, the resonances due to the wave's reflection from the practical boundary cannot be taken into account. The assumption of the PMC boundary is accurate for the computation of the impedance of the power–ground planes, but it is not suitable for the computation of the radiated field. This is because that the electromagnetic waves cannot penetrate the PMC boundary, that is, the internal and external electromagnetic fields are factitiously isolated. That is not true for real power–ground planes. One way to consider the radiation's effect on the impedance of power–ground planes is increasing the loss tangent of the substrate between the power and ground planes [15]. However, the assumption of a PMC boundary is still applied, which prevents an accurate prediction of the radiation.

In this section, we propose a 3D hybrid integral equation method. It is based on a power–ground plane's model, which differs from established models in that it considers the finite size and rectangular shape of the power–ground planes and does not assume a PMC boundary. Because we introduce this non-PMC boundary, we can accurately calculate the external radiated field that is produced by the source inside the power–ground planes, as well as the electric current inside the power–ground planes that is induced by external sources. Therefore, this proposed method is able to solve all aspects of the EMC problems related to the power–ground panes, such as the coupling, emission, and susceptibility problems. Moreover, in practical designs, the power–ground plane is often split into several "islands" to provide powers for different circuits. The gaps between these islands will introduce significant electromagnetic emissions. These gaps are also considered in our model. The proposed method is optimized by making full use of the structural features of the power–ground planes, so that it not only provides a much accurate solution but also reduces the total number of unknowns.

3.2.1 Formulation

A typical finite power–ground structure is shown in Figure 3.20, where the upper and lower planes are assumed to be perfect electric conductors. The planes can be split into several parts to supply powers to the different circuit modules. In Figure 3.20, E^{in} and E^{out} denote the potential internal sources and external sources, respectively. The objective of the study is to calculate the fields generated inside and outside of the power–ground planes by these sources. The detailed formulation is presented in the following text.

First, the entire domain is divided into the internal subdomain (including the substrate that is sandwiched between the power and ground planes) and the external subdomain (the free space surrounding the power–ground planes). Let S_m denote the periphery and gaps of the power–ground planes (the gray regions in Figure 3.20a), and S be the closed surface of the internal subdomain (including S_m and the upper and lower planes) with $S_m \in S$. After that, by using

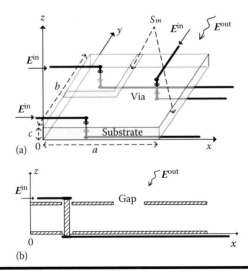

Figure 3.20 **A typical power–ground structure: (a) bird view and (b) cross section.**

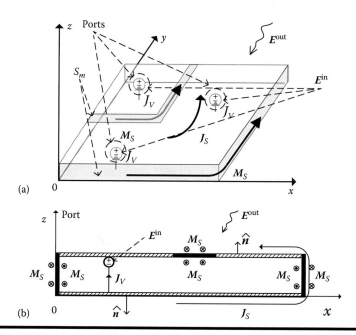

Figure 3.21 Equivalent problem of the power–ground planes: (a) bird view and (b) cross section.

the equivalence principle, S_m is closed by perfect electric conductors, and the equivalent magnetic currents M_S with the same values and opposite directions are placed on both sides of S_m. The equivalent electric current J_S is placed on the outside of S, and the vias are replaced by the equivalent volume electric currents J_V, as shown in Figure 3.21. Finally, each subdomain can be modeled by using the respective integral equations, which are coupled with the help of the equivalent currents M_S and J_S.

For the external subdomain, the integral kernel is the free-space Green's function, whereas for the internal subdomain, the integral kernel is the rectangular cavity's dyadic Green's function [19] with the following features:

1. There is no need for unknown electric currents to be placed on the inside of S, because their effects have already been taken care of by the dyadic Green's functions themselves.

2. When the thickness of the substrate is much smaller than the shortest wavelength of interest, only low-order modes are considered inside the power–ground planes. This is because that the high-order modes decay very quickly along their propagation paths. In this case, the dyadic Green's functions of the internal problem are reduced to 2D functions (while the external problem is still a 3D problem), which reduces the complexity of the internal problem.

For the internal subdomain, two integral equations are derived:
Due to the continuity of the tangential magnetic field on S_m, we get

$$\left[L^h\left[r, M_S\left(r'\right)\right] + K^h\left[r, \dot{J}_V\left(r'\right)\right]\right]_t + \hat{n}/2 \times \dot{J}_S\left(r\right) = 0 \qquad (3.21)$$

with the observation point $r \in S_m$.
Due to the vanishing of the total tangential electric field on vias surface, we get

$$\left[K^e\left[r, M_S\left(r'\right)\right] - L^e\left[r, \dot{J}_V\left(r'\right)\right]\right]_t = E_t^{in}\left(r\right) \qquad (3.22)$$

with the observation point r located on the vias.
For the external subdomain, one integral equation is derived:
Due to the vanishing of the total tangential electric field on the outside of S, we get

$$\left[K^0\left[r, M_S\left(r'\right)\right] + L^0\left[r, \dot{J}_S\left(r'\right)\right]\right]_t + \hat{n}/2 \times M_S\left(r\right) = E_t^{out}\left(r\right) \qquad (3.23)$$

with the observation point $r \in S$.
In Equations 3.21 through 3.23, the unknown functions to be solved for are M_S, $\dot{J}_V$, and $\dot{J}_S$. $\dot{J}_S = J_S \eta_0$ and $\dot{J}_V = J_V \eta_0$ denote the scaled equivalent electric currents on the outside of S and along vias with $\eta_0 = 120\pi$, respectively. Equations 3.21 through 3.23 are not independent of each other; they must be solved together to get the values of the unknowns. The common unknowns in internal and external equations are $\dot{J}_S$ and M_S. They are used to couple the internal and external subdomains. They are related to the surface electromagnetic fields E and H as follows:

$$\dot{J}_S\left(r\right) = \hat{n} \times \dot{H}\left(r\right) \text{ and } M_S\left(r\right) = E\left(r\right) \times \hat{n} \qquad (3.24)$$

where:
$\hat{n}$ denotes the outward unit normal vector of S, as shown in Figure 3.21b
$\dot{H} = H\eta_0$ denotes the scaled magnetic field
E_t^{in} denotes the internal excitation applied on the vias, which represents the SSN that comes from the integrated circuit connected to the vias
E_t^{out} denotes the incident wave from the outside of the power–ground planes
The subscript t denotes the tangential component

Similar to the feed modeling commonly used in method of moments (MoM) simulations of wire antennas, E_t^{in} is modeled as delta-gap voltage sources as shown in Figure 3.21. Ports are defined together with these delta-gap voltage sources, so that the power–ground planes with N vias are equivalent to an N-port network. Its admittance/impedance matrix can be calculated by solving $\dot{J}_V$.
The superscripts h, e, and 0 in the operators L and K denote the magnetic field in the cavity, the electric field in the cavity, and the electric field in free space,

respectively. The operators L^e, L^b, K^e, K^b, L^0, and K^0 in Equations 3.21 through 3.23 can be expressed in two different forms according to the different Green's functions used: One is to directly use the electric dyadic Green's functions of the first and second kinds $\bar{G}_{e1}$ and $\bar{G}_{e2}$ (this form represents the electric and magnetic field integral equations); another is to use the magnetic and electric vector potentials $\bar{G}_A$ and $\bar{G}_F$, as well as the electric and magnetic scalar potentials G_Φ and G_Ψ (this form represents the mixed potential integral equation). The later one has weaker singularities than the former one. Therefore, when discretizing L^e, L^b, K^e, K^b, L^0, and K^0 into linear equations, it is easier to accurately calculate the coefficient matrix by using the potential dyadic Green's functions.

It should be noted that the power–ground planes are highly resonant structures. At the resonance frequency, the numerical solution becomes sensitive to the coefficient matrix's values. According to these, the potentials with a weaker singularity are used to express the operators in Equations 3.21 through 3.23 in the following way:

$$L^e\left[r,X(r')\right]=\frac{jk_0\mu_r\displaystyle\int_V \bar{G}_A\left(r,r'\right)\cdot X\left(r'\right)dv' + j\nabla\displaystyle\int_V G_\Phi\left(r,r'\right)\nabla'\cdot X\left(r'\right)dv'}{k_0\varepsilon_r} \tag{3.25}$$

$$L^b\left[r,X(r')\right]=\frac{jk_0\varepsilon_r\displaystyle\int_S \bar{G}_F\left(r,r'\right)\cdot X\left(r'\right)ds' + j\nabla\displaystyle\int_S G_\Psi\left(r,r'\right)\nabla'_S\cdot X\left(r'\right)ds'}{k_0\mu_r} \tag{3.26}$$

$$K^b\left[r,X(r')\right]=\int_V \nabla\times\bar{G}_A\left(r,r'\right)\cdot X\left(r'\right)dv' \tag{3.27}$$

$$K^e\left[r,X(r')\right]=\int_S \nabla\times\bar{G}_F\left(r,r'\right)\cdot X\left(r'\right)ds' \tag{3.28}$$

$$L^0\left[r,X(r')\right]=\frac{jk_0\displaystyle\int_S G_0\left(r,r'\right)\cdot X\left(r'\right)ds' + j\nabla\displaystyle\int_S G_0\left(r,r'\right)\nabla'_S\cdot X\left(r'\right)ds'}{k_0} \tag{3.29}$$

$$K^0\left[r,X(r')\right]=\int_S \nabla G_0\left(r,r'\right)\times X\left(r'\right)ds' \tag{3.30}$$

where $X(r')$ is the unknown vector function ($\dot{J}_S$, $\dot{J}_V$, or M_S)

The prime on ds' and dv' emphasizes that the integration is over r'. $G_0(r,r')=(e^{-jk_0|r-r'|})/(4\pi|r-r'|)$ is the free-space Green's function with the free-space wave number $k_0 = \omega\sqrt{\varepsilon_0\mu_0}$, where ω is the angular frequency; ε_0 and μ_0 are the permittivity and permeability in vacuum, respectively; ε_r and μ_r are the relative permittivity and permeability of the substrate. The losses in substrate are accounted for by assuming complex ε_r and μ_r.

The relationships between potential Green's functions are $\nabla \cdot \bar{G}_A = -\nabla' G_\Phi$ and $\nabla \cdot \bar{G}_F = -\nabla' G_\Psi$. Under the Cartesian coordinate system defined in Figure 3.20, the expressions of these potential Green's functions are written as

$$\bar{G}_A(r,r') = \sum_{n=0}^{\infty}\sum_{m=0}^{\infty}\sum_{l=0}^{\infty} C_{mnl}$$

$$\begin{bmatrix} C_x S_y S_z C_x' S_y' S_z' & 0 & 0 \\ 0 & S_x C_y S_z S_x' C_y' S_z' & 0 \\ 0 & 0 & S_x S_y C_z S_x' S_y' C_z' \end{bmatrix} \tag{3.31}$$

$$\bar{G}_F(r,r') = \sum_{n=0}^{\infty}\sum_{m=0}^{\infty}\sum_{l=0}^{\infty} C_{mnl}$$

$$\begin{bmatrix} S_x C_y C_z S_x' C_y' C_z' & 0 & 0 \\ 0 & C_x S_y C_z C_x' S_y' C_z' & 0 \\ 0 & 0 & C_x C_y S_z C_x' C_y' S_z' \end{bmatrix} \tag{3.32}$$

$$G_\Phi = \sum_{n=0}^{\infty}\sum_{m=0}^{\infty}\sum_{l=0}^{\infty} C_{mnl} S_x S_y S_z S_x' S_y' S_z' \tag{3.33}$$

$$G_\Psi = \sum_{n=0}^{\infty}\sum_{m=0}^{\infty}\sum_{l=0}^{\infty} C_{mnl} C_x C_y C_z C_x' C_y' C_z' \tag{3.34}$$

where:

$$C_{mnl} = \frac{\delta_m \delta_n \delta_l}{abc\left(k_x^2 + k_y^2 + k_z^2 - k_0^2 \varepsilon_r \mu_r\right)}$$

$$\delta_m = \begin{cases} 1, & m = 0 \\ 2, & m \neq 0 \end{cases}, \quad \delta_n = \begin{cases} 1, & n = 0 \\ 2, & n \neq 0 \end{cases}, \quad \text{and} \quad \delta_l = \begin{cases} 1, & l = 0 \\ 2, & l \neq 0 \end{cases}, \quad \text{for } m, n, l = 0,1,2,\ldots$$

$k_x = m\pi/a$, $k_y = n\pi/b$ and $k_z = l/c$, $C_\beta = \cos(k_\beta \beta)$, $S_\beta = \sin(k_\beta \beta)$, $C_\beta' = \cos(k_\beta \beta')$, and $S_\beta' = \sin(k_\beta \beta')$, for $\beta = x/y/z$ and $\beta' = x'/y'/z'$. The length, width, and height of the power–ground planes are defined to be a, b, and c, respectively, as shown in Figure 3.20a.

After the integral Equations 3.21 through 3.23 are obtained, we need to discretize them into the linear equations, which can be solved by the computer. To do so, first, the unknown currents J_S and M_S are expanded by using the subsectional vector basis function defined in [20], whereas $\dot{J}_V$ is expanded by using the subsectional vector basis function which is uniform along each via, where each via is modeled as the cylinder with finite radius. After that, the basis functions of M_S are used to perform the dot products on both sides of integral Equation 3.21. Subsequently, these dot products are integrated over each basis function's domain. This is the so-called Galerkin's process. The similar Galerkin's process is applied between the basis functions of $\dot{J}_V$ and integral Equation 3.22, and the basis functions of $\dot{J}_S$ and the integral Equation 3.23. After this discretization, the variables r and r' in Equations 3.21 through 3.30 vanish and the following linear equations can be obtained:

The discretization of the internal integral Equations 3.21 and 3.22 leads to

$$[A]\begin{bmatrix} M_S \\ \dot{J}_V \end{bmatrix}+[C]\begin{bmatrix} \dot{J}_S \end{bmatrix}=\begin{bmatrix} E_t^{\text{in}} \end{bmatrix} \tag{3.35}$$

The external integral Equation 3.23 is discretized to

$$[D]\begin{bmatrix} M_S \\ \dot{J}_V \end{bmatrix}+[Z]\begin{bmatrix} \dot{J}_S \end{bmatrix}=\begin{bmatrix} E_t^{\text{out}} \end{bmatrix} \tag{3.36}$$

where $[M_S]$, $[\dot{J}_V]$, and $[\dot{J}_S]$ denote the vectors of expansion coefficients of the unknown functions. The coefficients matrices $[A]$, $[C]$, $[D]$, and $[Z]$ are obtained through the discretization of the integral operators L^e, L^h, K^e, K^h, L^0, K^0, and $\hat{n}/2\times$ in Equations 3.21 through 3.23. These operators describe the electromagnetic interactions between M_S, $\dot{J}_V$, and $\dot{J}_S$. Because the Galerkin's process is used, matrices $[A]$ and $[Z]$ are symmetric. It should be noted that matrices $[D]$, $[Z]$, and $[C]$ are independent of the vias' layout, so that they can be reused for the same power–ground planes with different vias layouts. Either Equation 3.35 or 3.36 is underdetermined; they must be solved together to get the values of the equivalent currents M_S and $\dot{J}_S$. This means that M_S and $\dot{J}_S$ are dependent on both the internal and external geometries, as opposed to merely one of them. However, we can see from Equations 3.35 and 3.36 that both the internal source $[E_t^{\text{in}}]$ and the external source $[E_t^{\text{out}}]$ may contribute to M_S and $\dot{J}_S$.

The linear Equations 3.35 and 3.36 are often iteratively solved for a larger number of unknowns. However, the convergence will be very slow at the resonance frequencies of the power–ground planes due to the ill-conditioned matrix $[A]$. In this section, we use the same method used in [21] to perform the LU factorization of $[A]$, where the symmetry of the matrix $[A]$ is used to accelerate the calculation. After that, we can decouple Equations 3.35 and 3.36, and solve the following equivalent equations:

$$\left([Z]-[D]\cdot[A]^{-1}\cdot[C]\right)[\dot{J}_S]=[E_t^{\text{out}}]-[D]\cdot[A]^{-1}\cdot[E_t^{\text{in}}] \tag{3.37}$$

$$\begin{bmatrix} M_S \\ \dot{J}_V \end{bmatrix} = [A]^{-1} \cdot \left(\begin{bmatrix} E_t^{in} \end{bmatrix} - [C] \cdot \begin{bmatrix} \dot{J}_S \end{bmatrix} \right) \tag{3.38}$$

Equation 3.37 is iteratively solved for $\dot{J}_S$ by using the method in [22,23]; then M_S and $\dot{J}_V$ can be calculated from Equation 3.38 by using the previously obtained $\dot{J}_S$. Subsequently, the impedance matrix of the power–ground planes and the induced electric current along the vias can be calculated by using $\dot{J}_V$, whereas the external radiated field can be calculated by using $\dot{J}_S$ and M_S. In general, the impedance, the induced current, and the radiated field are expected to be as low as possible to ensure that the high-speed electronic devices work at a normal condition.

Because the radiation from the power–ground planes is considered in the proposed method, the calculated impedance matrix is not an ideal lossless network matrix even for lossless substrates. This is different from the most available methods where the PMC boundary assumption is applied. In Section 3.2.2, the proposed method is validated through several numerical examples. Because both the internal and external sources are considered in the proposed method, it can be used for both emission and susceptibility simulations.

3.2.2 Validation and Discussion

For validation purpose, the proposed method is compared to measurement data, as well as to results from the commercial simulators FEKO [24] and HFSS [16].

3.2.2.1 Input and Mutual Impedances of Power–Ground Planes

A single-layer power–ground structure with one via is shown in Figure 3.22. The radius of the via is 350 μm, and the substrate material is FR4 with a thickness of 0.1588 cm, $\varepsilon_r = 4.1$, and $\mu_r = 1$. In Figure 3.23, the input impedance of the via is calculated by using the proposed method, and compared to measurement results from [15]. Four modes are excited according to the dimension of the power–ground planes and the location of the via: TM_{10}, TM_{20}, TM_{30}, and TM_{02}, where TM_{30} and TM_{02} are degenerated. Good agreement can be observed.

The second example is a single-layer power–ground structure with two vias as shown in Figure 3.24. The substrate material is FR4 with a 0.6 mm thickness, a

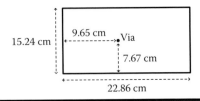

Figure 3.22 Power–ground planes with one via.

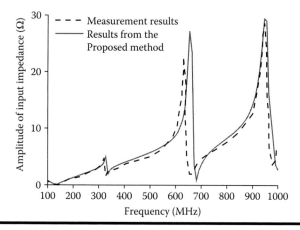

Figure 3.23 **Magnitude of the input impedance of the via in Figure 3.22.**

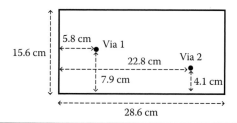

Figure 3.24 **Power–ground planes with two vias.**

relative permittivity of 4.2, and a loss tangent of 0.018. The radius of two vias is 100 μm. Figure 3.25 shows the calculated mutual impedance between two vias by using the proposed method along with measurement results from [25]. From this figure, we can see that the agreement is good for the low-order modes (of which the resonance frequencies are below 1.5 GHz), whereas for the high-order modes, the measured resonance frequencies are slightly higher than the calculated ones. This difference is due to measurement uncertainties. At high frequencies, the permittivity and loss tangent of the substrate often differ from their nominal values at low frequencies. And the measured S parameters are more sensitive to the parasitic capacitance and inductance than those at low frequencies. These affect the high-order modes' resonance frequencies. The small shifts of the measured resonance frequencies of high-order modes are also observed in Figure 3.22 of [25].

3.2.2.2 Use of Shorting Pins to Reduce Radiation from Gaps

In practical package design, the power–ground plane is often split into several parts to supply powers for different circuit modules. The gaps between different parts

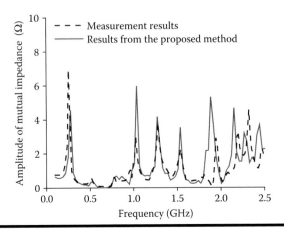

Figure 3.25 Magnitude of the mutual impedance between two vias shown in Figure 3.24.

usually degrade the shielding performance of the power–ground plane, that is, they strengthen the leakage radiation, compared to the unsplit power–ground plane. The next example is used to validate the ability of the proposed method to model the interactions between internal sources and external radiated fields, as well as to exam the use of shorting pins to reduce the gaps' radiation. The dimensions of the power–ground planes are plotted in Figure 3.26. The substrate material is of a 0.2 mm thickness, $\varepsilon_r = 1$, and $\mu_r = 1$. The radius of the active via and shorting pins is 50 μm. The upper plane is split into two parts at $x = 6$ mm, and an active via is placed at the center of the left part. To reduce the radiated field from the gap, four shorting pins are evenly distributed along the gap as shown in Figure 3.26.

The electric shielding effectiveness (SE) is a common EMC measure used in EMC engineering to evaluate the shielding performance. In this section, it is defined as the ratio of the radiated field without the shorting pins to the radiated

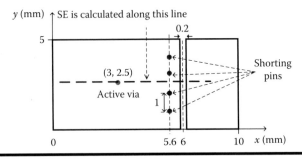

Figure 3.26 A split power–ground structure with four shorting pins (unit: mm). The SE is calculated along the line: from (1, 2.5, and 0.4 mm) to (8, 2.5, and 0.4 mm).

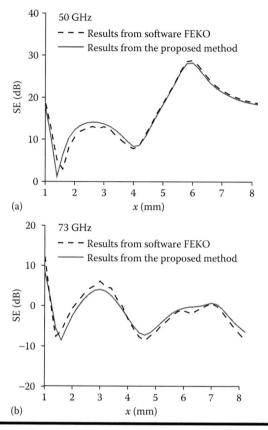

Figure 3.27 **Electric SE of the shorting pins at (a) 50 GHz and (b) 73 GHz, along the line from (1, 2.5, and 0.4 mm) to (8, 2.5, and 0.4 mm).**

field with the presence of the shorting pins, measured at the same line, which is located 0.2 mm above the upper plane in Figure 3.26.

Figure 3.27a,b shows the electric SE of the shorting pins at 50 and 73 GHz, respectively. The lowest resonance frequency of the left part of the power–ground planes is at 50 GHz according to its dimension and the location of the active via. At this resonance frequency, the radiated field from the gap is very strong. It can be observed from Figure 3.27a that by using the shorting pins, this radiated field can be reduced by 10 dB for most observation points. However, at 73 GHz, we can see from Figure 3.27b that for most observation points, the SE is negative, that is, the use of the shorting pins results in an even higher radiation. This is due to a resonance between the shorting pins. They divide the whole gap into five equivalent magnetic currents, with 73 GHz being close to their resonance frequency. This suggests that for the practical package design, the distance between shorting pins should be carefully chosen according to the operating frequencies.

In order to validate the accuracy of the proposed method to model the interaction between the external radiated field and the internal source, the simulations results are compared to results obtained from the commercial software FEKO and are plotted in Figure 3.27. Again, a good agreement is observed.

3.2.2.3 Induced Electric Current due to External Noise

In the above examinations, the sources are located inside the power–ground planes, the coupling inside the power–ground planes and radiated fields are calculated. These correspond to the conducted and radiated emission issues in EMC engineering. In the following examination, the source is located outside the power–ground planes, and the induced electric currents along vias inside the power–ground planes are calculated. This corresponds to another important EMC issue: the susceptibility. For the power–ground planes, the external source can induce electric currents along the vias, which will then propagate and produce noise similar to SSN. These induced currents are usually small, but when their spectrums cover any resonant frequency of the power–ground planes, they will be amplified.

Figure 3.28 shows the considered power–ground planes. The upper plane is split into two parts and each part has one via located inside. The substrate between the upper and lower planes is made of FR4 with a thickness of 0.1 mm, a relative permittivity of 4.4, and a loss tangent of 0.02. The radius of two vias is 0.1 mm. In this simulation, for the general case, the external source is assumed to be a plane wave with $-x$ polarization and $-z$ incident direction, and its amplitude is 1 V/m, as shown in Figure 3.28.

Figure 3.29a,b shows the induced electric currents along via 1 and via 2, respectively. The gap on the upper plane degrades the shielding of the power–ground planes by increasing the leakage of the electric field into the structure. The induced electric currents are amplified at the resonant frequencies of the power–ground

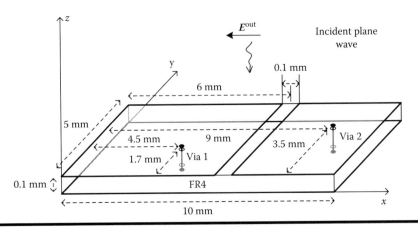

Figure 3.28 A power–ground structure and the incident plane wave.

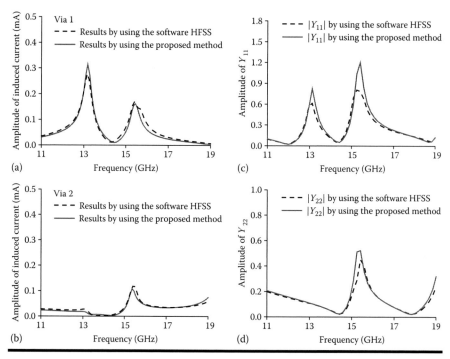

Figure 3.29 Simulation results for (a) the amplitude of the induced current along via 1, (b) the amplitude of the induced current along via 2, (c) the input admittance of via 1, and (d) the input admittance of via 2.

planes. Figure 3.29c,d shows the calculated input admittances of via 1 and via 2, respectively. We can observe from Figure 3.29a,c as well as from Figure 3.29b,d respectively, that the resonant frequencies of the induced currents are the same as those of the input admittances. This is because the resonant frequencies of a power–ground structure are inherently determined by its geometry and via layout, that is, the coefficients matrices $[A]$, $[C]$, $[D]$, and $[Z]$ in Equations 3.35 and 3.36, whereas the excitations $[E_t^{in}]$ and $[E_t^{out}]$ at the right-hand sides of Equations 3.35 and 3.36 only affect the amplitudes of these resonances.

For validation purpose, the results from the commercial simulator HFSS [16] are also plotted in Figure 3.29. Again, a good agreement can be observed. However, the proposed method outperforms HFSS in terms of CPU time (37 minutes as opposed to two hours and 20 minutes).

3.2.3 Conclusion

In this section, a hybrid 3D integral equation method is proposed to solve the electromagnetic emission and susceptibility of power–ground planes with a finite size. Through numerous numerical examinations, the accuracy and efficiency of this method are validated. The major advantage of this method is its ability to simulate coupling, emission, and susceptibility problems related to the power–ground planes at the same time.

3.3 Power–Ground Planes with Narrow Slots

Several approaches have been proposed to minimize the power–ground plane noise [26,29]. The three most frequently used methods are decoupling capacitors, isolation, and electromagnetic bandgap. The decoupling capacitor is band-limit due to its series inductance. The isolation and planar electromagnetic bandgap approaches make use of narrow slots etched on the power or ground planes to isolate the noisy circuits from other sensitive circuits, and to prevent the propagation of power–ground noise.

However, the slots on the power–ground planes also produce undesired effects on signal and power integrity. Efficient and accurate electromagnetic simulation algorithms are required for analyzing the power–ground planes with narrow slots. A combined circuit method is proposed in [30,32], where the pair of power–ground planes is simulated by either an analytical [1] or a numerical method [33], and slots are modeled as two coupled transverse electromagnetic transmission lines. The width of the transmission line is experimentally chosen. Different slot widths result in different values of coupling capacitance across the slot. The TEM approximation of the slots is not accurate at high frequencies. More accurate modeling can be achieved by employing 3D full-wave methods, which include the PEEC method [4], the finite difference method [6], and the MoM [10]. However, these techniques are, in general, expensive in terms of CPU and memory requirements.

A better solution for this complex problem is to simplify the full-wave method by considering the modal field distribution in the space between the power and ground planes. In Section 3.1, 2D IEEC method [34] has been introduced to do this. However, the slots in the power–ground planes are not considered yet. In this section, the electromagnetic coupling across slots is accurately calculated. Because the high-order modes decay quickly in the direction of propagation, only the fundamental mode is considered when computing the interaction between the slot and the power–ground planes. In this way, the complex surface integrals are reduced to line integrals, and the computing time is significantly reduced while maintaining good accuracy.

3.3.1 Formulation

3.3.1.1 Line Integral Equations

Figure 3.30 shows a typical pair of power–ground planes with a slot and a via clearance hole. The via clearance (also called an "antipad" in some publications) is the void between the via and the power–ground planes. The via is used to connect

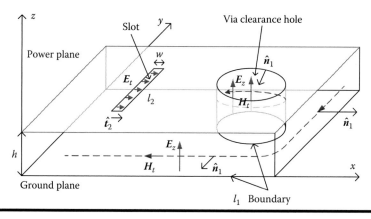

Figure 3.30   **Finite-size power–ground planes with narrow slots.**

signal traces situated in different layers. Usually, the thickness of the substrate sandwiched between the power and ground planes is much smaller than the wavelength of interest, so that the electromagnetic fields along the boundary of the power–ground planes and via clearance hole are assumed to be uniform along the z-direction. In Figure 3.30, E_z and H_t denote the tangential electric and magnetic fields along the boundaries of power–ground planes and via clearance holes, respectively; l_1 is the contour of this boundary in the xy plane with $\hat{n}_1$ as its outward normal vector. E_t is the tangential electric field across the slot surface, polarized in $\hat{t}_2$-direction; l_2 is the axis of the slot.

By using the equivalence principle, the slot and via clearance are closed, and the original *finite-size* power–ground planes are extended to *infinite* parallel planes. All discontinuities, including the slot, via clearance, and boundary, contribute to the conduction and emission noises, which can be analyzed by replacing the discontinuities with equivalent electric and magnetic currents as appropriate. We place the equivalent electric current J and magnetic current M_1 on the boundaries of the power–ground planes and via clearance, and the magnetic current M_2 on the slot as shown in Figure 3.31.

$$M_1 \equiv \hat{n}_1 \times E_z, \; J \equiv H_t \times \hat{n}_1 \text{ and } M_2 \equiv \hat{z} \times E_t \qquad (3.39)$$

We also define the corresponding voltages and currents as

$$V_1 \equiv -hE_z, V_2 \equiv wE_t \text{ and } J = \hat{z}\,J \qquad (3.40)$$

where:
   h is the height of the substrate
   w is the width of the slot

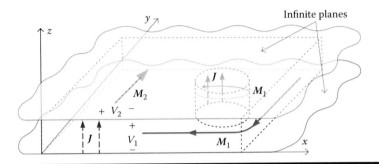

Figure 3.31 **Equivalent infinite power–ground planes with equivalent electric and magnetic currents.**

The relationship between (M_1, M_2) and (V_1, V_2) is

$$M_1 = \frac{-\hat{l}_1 V_1}{h} \quad \text{and} \quad M_2 = \frac{-\hat{l}_2 V_2}{w} \tag{3.41}$$

with $\hat{l}_1 \equiv \hat{n}_1 \times \hat{z}$ and $\hat{l}_2 \equiv \hat{t}_2 \times \hat{z}$. Because the Green's functions of the infinite parallel planes are available, the problem of Figure 3.31 can be efficiently solved by using the integral equation method.

The magnetic vector potential $\bar{G}_A$, electric vector potential $\bar{G}_F$, electric scalar potential G_q, and magnetic scalar potential G_p of the infinite parallel planes pair can be written as

$$\bar{G}_A = G_q(\hat{x}\hat{x} + \hat{y}\hat{y}) + G_p \hat{z}\hat{z} \tag{3.42}$$

$$\bar{G}_F = G_p(\hat{x}\hat{x} + \hat{y}\hat{y}) + G_q \hat{z}\hat{z} \tag{3.43}$$

$$G_q = \frac{1}{j4h} \sum_{m=0}^{\infty} \delta_m H_0^{(2)}(k_\rho \rho) \sin(k_z z) \sin(k_z z') \tag{3.44}$$

$$G_p = \frac{1}{j4h} \sum_{m=0}^{\infty} \delta_m H_0^{(2)}(k_\rho \rho) \cos(k_z z) \cos(k_z z') \tag{3.45}$$

where:
$$\delta_m = \begin{cases} 1, & m = 0 \\ 2, & m \neq 0 \end{cases}, \; k_z = m\pi/h, \; \rho = \sqrt{(x-x')^2 + (y-y')^2} \text{ and } k_\rho = \sqrt{k^2 - k_z^2}$$

k is the wavenumber in the substrate
$H_0^{(2)}$ is the zero-order Hankel function of the second kind

Green's functions Equations 3.42 through 3.45 include all modal fields between the infinite parallel planes. For high-order modes with $m \neq 0$, because h is much smaller than the wavelength of interest, k_ρ is approximated as a larger imaginary number. Therefore, high-order modes decay very quickly and contribute only to the local electromagnetic coupling.

In terms of the Green's functions in Equations 3.42 through 3.45, the electric and magnetic fields produced by the electric current J and magnetic current M are

$$E(r) = L^e\left[J(r')\right] + K^e\left[M(r')\right] \tag{3.46}$$

$$H(r) = K^h\left[J(r')\right] + L^h\left[M(r')\right] \tag{3.47}$$

where the operators are defined as

$$L^e\left[J(r')\right] = -j\omega\mu\int_S \bar{G}_A(r,r')\cdot J(r')ds' - \frac{j}{\omega\varepsilon}\nabla\int_S G_q(r,r')\nabla'_s\cdot J(r')ds' \tag{3.48}$$

$$K^e\left[M(r')\right] = -\int_S \nabla\times\bar{G}_F(r,r')\cdot M(r')ds' \tag{3.49}$$

$$L^h\left[M(r')\right] = -j\omega\varepsilon\int_S \bar{G}_F(r,r')\cdot M(r')ds' - \frac{j}{\omega\mu}\nabla\int_S G_p(r,r')\nabla'_s\cdot M(r')ds' \tag{3.50}$$

$$K^h\left[J(r')\right] = \int_S \nabla\times\bar{G}_A(r,r')\cdot J(r')ds'. \tag{3.51}$$

Superscripts e and h denote the electric and magnetic fields, respectively. ε and μ are the permittivity and permeability of the substrate, respectively.

Due to the continuity of the tangential electric field on the boundaries of the power–ground planes and via clearance holes, the electric field integral equation becomes

$$K^e_z\left[M_1(r')\right] + K^e_z\left[M_2(r')\right] + L^e_z\left[J(r')\right] = \frac{E_z(r)}{2} \tag{3.52}$$

where $r \in l_1$, subscript z indicates the z-component

Due to continuity of the tangential magnetic field on the slot, another magnetic field integral equation can be written as

$$L^h_t\left[M_1(r')\right] + L^h_t\left[M_2(r')\right] + K^h_t\left[J(r')\right] = -2L^0_t\left[M_2(r')\right] \tag{3.53}$$

Where $r \in l_2$, subscript t indicates the component tangential to the slot surface. $2L_t^0[M_2(r')]$ is the tangential magnetic field produced by M_2 placed on an infinite ground plane in free space

$$L^0[M_2(r')] = -j\omega\varepsilon_0 \int_S G_0(r,r') \cdot M_2(r')ds'$$

$$-\frac{j}{\omega\mu_0} \nabla \int_S G_0(r,r')\nabla'_s \cdot M_2(r')ds' \tag{3.54}$$

Where $G_0(r,r') = e^{-jk_0\rho}/4\pi\rho$ is the free-space Green's function and ε_0, μ_0, and k_0 are the permittivity, permeability, and wavenumber of the free space, respectively.

Equations 3.52 and 3.53 are written as

$$K_z^e[M_1(r')] - E_z(r)/2 + K_z^e[M_2(r')] = -L_z^e[J(r')] \tag{3.55}$$

$$L_t^h[M_1(r')] + (L_t^h + 2L_t^0)[M_2(r')] = -K_t^h[J(r')] \tag{3.56}$$

The above equations involve complex surface integrals. The direct solutions of these equations are time-consuming. In the following, we will simplify these equations to only involve line integrals.

Performing the integration $\int_0^h dz$ on both sides of Equation 3.55, we get

$$\frac{k}{2j} \int_{l_1} H_1^{(2)}(k\rho)\hat{\rho} \cdot \hat{n}_1 V_1(r')dl' + V_1(r) + \frac{k}{2j} \int_{l_2} H_1^{(2)}(k\rho)\hat{\rho} \cdot \hat{t}_2 V_2(r')dl'$$

$$= \frac{\omega\mu h}{2} \int_{l_1} H_0^{(2)}(k\rho)J(r')dl' \quad \text{with } r \in l_1 \tag{3.57}$$

By taking the dot product of both sides of Equation 3.56 with $\hat{l}_2$, we get

$$\hat{l}_2 \cdot \frac{\omega\varepsilon}{4h} \int_{l_1} H_0^{(2)}(k\rho)V_1(r')dl' + \frac{1}{4\omega\mu h} \frac{\partial}{\partial l_2} \int_{l_1} H_0^{(2)}(k\rho)\frac{\partial V_1(r')}{\partial l'}dl'$$

$$+ \hat{l}_2 \cdot j\omega\varepsilon \int_{l_2} \left(G_p + \frac{2G_0}{\varepsilon_r}\right)\bigg|_{z=z'=h} V_2(r')dl'$$

$$+ \frac{j}{\omega\mu} \frac{\partial}{\partial l_2} \int_{l_2} (G_p + \mu_r 2G_0)\bigg|_{z=z'=h} \frac{\partial V_2(r')}{\partial l'}dl' \tag{3.58}$$

$$= \frac{k\hat{t}_2 \cdot}{4j} \int_{l_1} H_1^{(2)}(k\rho)J(r')\hat{\rho}dl' \quad \text{with } r \in l_2$$

In Equations 3.57 and 3.58, the relationship (Equation 3.41) is used, and M_1 and M_2 are replaced by V_1 and V_2, respectively. Because the width of the narrow slot is very small (usually as small as 1/100 of the interested wavelength), the magnetic current M_2 is assumed to flow along the axis of the slot. $H_1^{(2)}$ is the first-order Hankel function of the second kind. $\hat{\rho} = \left[(x - x')\hat{x} + (y - y')\hat{y} \right]/\rho$.

The following formula is used to derive Equations 3.57 and 3.58:

$$\int_0^b G_p dz = \int_0^b G_p dz' = \frac{H_0^{(2)}(k\rho)}{4j} \tag{3.59}$$

After applying the above equation, both Equations 3.57 and 3.58 are independent of z, and the original surface integrals are reduced to simple line integrals along the curves l_1 and l_2.

Because V_1 and J are assumed to be independent of z, Equations 3.48 through 3.51 and 3.59 show that the high-order modes ($m \neq 0$) do not contribute to the electromagnetic fields produced by V_1 and J. Therefore, only the fundamental mode ($m = 0$) is considered when computing the interactions between the slot, boundary, and via clearance. The Hankel functions $H_0^{(2)}$ and $H_1^{(2)}$ in Equations 3.57 and 3.58 represent the fundamental mode contribution. This saves considerable CPU time. The only high-order modes left to evaluate are

$$\left(G_p + 2\alpha G_0 \right)\Big|_{z=z'=b} = \frac{1}{j4h} \sum_{m=0}^{\infty} \delta_m H_0^{(2)}(k_\rho \rho) + \frac{\alpha e^{-jk_0\rho}}{2\pi\rho} \tag{3.60}$$

where $\alpha = 1/\varepsilon_r$ or μ_r. ε_r and μ_r are the relative permittivity and permeability of the substrate, respectively. Equation 3.60 is a quickly decaying series. The high-order modes in the equation account for the effects of the parasitic capacitance and inductance of the slot, so that the reduced integral Equations 3.57 and 3.58 still accurately model the slot.

In [10], a hybrid integral equation method is proposed to solve the power–ground planes with slots, where the Green's functions of a rectangular cavity and surface integrals are employed. The line integrals Equations 3.57 and 3.58 are more efficient in comparison with that method. When there is no slot, that is, $V_2 = 0$, Equation 3.57 is reduced to the Equation (1) in [34]. Therefore, [34] can be considered as a special case of the proposed method.

3.3.1.2 Solution of the Line Integral Equations

The line integral Equations 3.57 and 3.58 are converted into linear equations by using pulse basic functions in combination with the point matching method [35]. Similar to the procedure in Section 1.2.5, the curves l_1 and l_2 are divided into small segments, which are named as *current segments*. V_1, V_2, and J are assumed to

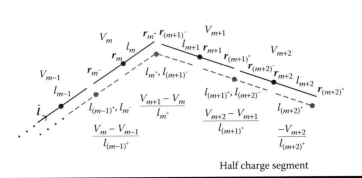

Figure 3.32 Current segments (solid lines) used for the expansion of V_1, V_2, and J, and charge segments (dashed lines) used for the expansion of $\partial V_1/\partial l$ and $\partial V_2/\partial l$.

be constants on each current segment. Between two adjacent current segments, a *charge segment* is defined. $\partial V_1/\partial l$ and $\partial V_2/\partial l$ are approximated by using the central difference and assumed to be constants on each charge segment. The current segments are used to expand the magnetic and electric current densities; the charge segments are used to expand the magnetic charge density. The current and charge segments are shown in Figure 3.32. r_{m^+} and r_{m^-} denote the front and back nodes of the *m*th current segment, respectively. l_m is the length of the *m*th current segment, and l_{m^+} or $l_{(m+1)^-}$ is the length of the charge segment.

The segment-based expansions of V_1, V_2, J, $\partial V_1/\partial l$, and $\partial V_2/\partial l$ are substituted into Equations 3.57 and 3.58. By enforcing Equations 3.57 and 3.58 at the center of each current segment, we can get the following linear equations:

$$\begin{bmatrix} U_{11} & U_{12} \\ U_{21} & U_{22} \end{bmatrix} \cdot \begin{bmatrix} V_1 \\ V_2 \end{bmatrix} = \begin{bmatrix} H_1 \\ H_2 \end{bmatrix} [I] \tag{3.61}$$

where V_1, V_2, and I are the expansion coefficients of V_1, V_2, and J, respectively.

The elements in submatrices U_{11}, U_{12}, U_{21}, U_{22}, H_1, and H_2 are calculated as

$$U_{m,n}^{(11)} = \frac{k l_n}{2j} F(m,n) \cdot \hat{\boldsymbol{n}}_{1n} + \delta_{m,n} \tag{3.62}$$

$$U_{m,n}^{(12)} = \frac{k l_n}{2j} F(m,n) \cdot \hat{\boldsymbol{t}}_{2n} \tag{3.63}$$

$$U_{m,n}^{(21)} = \frac{\omega\varepsilon}{4h} \Psi_1(m,n) \boldsymbol{l}_m \cdot \boldsymbol{l}_n$$

$$+ \frac{1}{4\omega\mu h} \left[\Psi_1\left(m^+,n^-\right) - \Psi_1\left(m^+,n^+\right) - \Psi_1\left(m^-,n^-\right) + \Psi_1\left(m^-,n^+\right) \right] \tag{3.64}$$

$$U_{m,n}^{(22)} = j\omega\varepsilon\Psi_2\left(m,n,1/\varepsilon_r\right)\boldsymbol{l}_m \cdot \boldsymbol{l}_n$$

$$+ \frac{j}{\omega\mu}\left[\Psi_2\left(m^+,n^-,\mu_r\right) - \Psi_2\left(m^+,n^+,\mu_r\right) - \Psi_2\left(m^-,n^-,\mu_r\right) + \Psi_2\left(m^-,n^+,\mu_r\right)\right]$$

(3.65)

$$H_{m,n}^{(1)} = \frac{\omega\mu b}{2}\Psi_1\left(m,n\right)$$

(3.66)

$$H_{m,n}^{(2)} = \frac{kl_m}{4j}\boldsymbol{F}\left(m,n\right) \cdot \hat{\boldsymbol{t}}_{2m}$$

(3.67)

where $\delta_{m,n} = \begin{cases} 1, & m = n \\ 0, & m \neq n \end{cases}$

Functions $\Psi_1\left(m,n\right)$, $\Psi_2\left(m,n,\alpha\right)$, and $\boldsymbol{F}\left(m,n\right)$ are defined as

$$\Psi_1\left(m,n\right) = \frac{1}{l_n}\int_{l_n} H_0^{(2)}\left(k\rho\right)dl'$$

(3.68)

$$\Psi_2\left(m,n,\alpha\right) = \frac{1}{l_n}\int_{l_n} \left(G_p + 2\alpha G_0\right)\Big|_{z=z'=b} dl'$$

(3.69)

$$\boldsymbol{F}\left(m,n\right) = \frac{1}{l_n}\int_{l_n} H_1^{(2)}\left(k\rho\right)\hat{\rho}\, dl'$$

(3.70)

The integral kernels in Equations 3.68 through 3.70 are singular when the testing point falls on the source segment. In that case, the kernel is broken into a singular part and a nonsingular part. The nonsingular part is integrated numerically, whereas the singular part is integrated analytically by using the method described in the appendix [36].

An equivalent network of the power–ground planes with slots can be derived if we define ports on each current segment along l_1 and l_2. V_1, V_2, and $\boldsymbol{I}$ in Equation 3.61 are then taken as the port voltages and currents.

1. Because the thickness of the substrate is much smaller than the wavelength, the perfect magnetic boundary condition is assumed along the boundary of the power–ground planes. Therefore, $\boldsymbol{I}$ is zero on the ports of this boundary.
2. The ports along each via clearance are connected in parallel to form one new port.

Substituting the above boundary conditions into Equation 3.61, we get a reduced equivalent network, where the new ports are defined on each via clearance.

This equivalent network represents the effects of power–ground planes with slot on the return currents of signal traces passing through the same pair of power–ground planes. It can be integrated with the equivalent circuits of the vias and signal traces [34] for signal and power integrity cosimulations.

3.3.2 Validation and Discussion

3.3.2.1 A Circular–Rectangular Shaped Pair of Power–Ground Planes with a Straight Slot

The first example is a circular–rectangular shaped pair of power–ground planes as shown in Figure 3.33, where a straight slot is etched on the upper plane. The substrate has a relative permittivity of 4.2, a loss tangent of $\tan\delta = 0.02$, and a thickness of 1.6 mm. Two SMA connectors are mounted on the upper planes. The inner conductors of the SMAs go through the via clearance holes and are soldered to the lower plane. The outer conductors of the SMAs are soldered to the upper plane. The network analyzer is connected to the SMAs to measure the S-parameters. The radius of the via clearance is 0.5 mm.

Figure 3.34 shows $|S_{11}|$ obtained by using the proposed IEEC method (denoted by IEEC), as well as measurements, and results computed with HFSS software[*] (which is based on the 3D finite element method). Good agreement can be observed in the figure. For this example, the proposed method yields results within one minute, whereas HFSS takes one hour. It should be noted that in the proposed method,

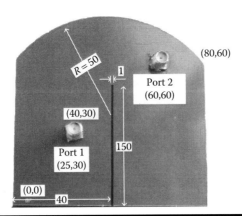

Figure 3.33 Circular–rectangular shaped pair of power–ground planes with a straight slot (unit: mm).

---

[*] Ansoft Software, Pittsburgh, PA. (Online). Available at: http://www.ansoft.com.

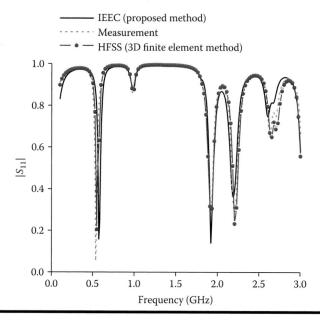

Figure 3.34 $|S_{11}|$ of the pair of power–ground planes of Figure 3.33.

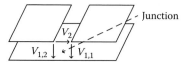

Figure 3.35 V_1 and V_2 at the junction between the slot and the boundary of the upper plane.

V_1 and V_2 follow Kirchhoff's loop rule at the junction between the slot and the boundary of the upper plane as shown in Figure 3.35. In the figure, $V_2 = V_{1,2} - V_{1,1}$.

Several combined field–circuit methods have been proposed to simulate the power–ground planes with slots [30,32], where the slot is modeled by two coupled TEM transmission lines. This TEM slot can be modeled in the proposed method by setting $L_t^0 = 0$ on the right-hand side of Equation 3.53. However, this TEM slot assumption is only accurate for low frequencies. Figure 3.36 shows the $|S_{11}|$ simulated with the software HFSS, where the magnetic field component tangential to the slot surface is set to zero. Also plotted are the $|S_{11}|$ curves obtained by the proposed method and by measurement. This figure shows that for high frequencies, the proposed method gives a more accurate result than the TEM slot assumption.

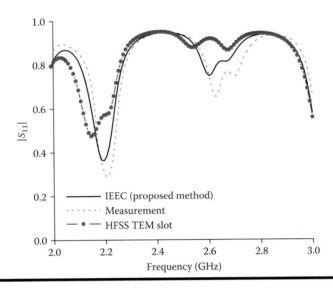

Figure 3.36 **Comparison between the proposed method and the TEM transmission line method.**

3.3.2.2 A Rectangular Shaped Pair of Power–Ground Planes with an Island

The second example is a rectangular-shaped pair of power–ground planes with an etched island on the upper plane as shown in the inset of Figure 3.37. The substrate has a relative permittivity of 4.1, a loss tangent $\tan\delta = 0.015$, and a thickness of 2 mm. The radius of the via clearance is 0.5 mm. Two ports are defined at the two via clearances. Figure 3.37 shows the magnitude of S_{21} calculated with the proposed method and the software SIWave (which is based on the 2D finite element method). The result obtained with the 3D software HFSS is also plotted for validation. This figure shows that the proposed method offers more accurate results than the SIWave software, especially at high frequencies. This is because the SIWave does not accurately consider the effects of the narrow slot. For this example, HFSS takes two hours, whereas both the proposed method and SIWave take one minute of computing time.

3.3.3 Conclusion

In this section, we discuss an efficient integral equation method for the simulation of power–ground planes with narrow slots. The integral equation solution has unique features in comparison with other available methods. It employs Green's functions of infinite parallel planes, and the unknowns are only defined on the

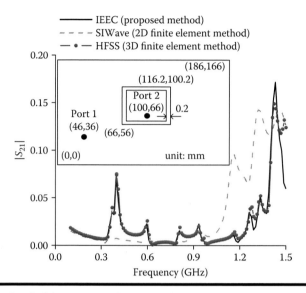

Figure 3.37 $|S_{21}|$ **of the rectangular pair of power–ground planes.**

discontinuities of the finite-size power–ground planes. In addition, considering the small separation between the power–ground planes, only the fundamental mode in the expressions of Green's functions is employed when computing the interactions between discontinuities. In this way, this method gives a highly efficient and accurate solution of complex 3D power–ground planes.

Appendix

The elements in $[U]$ and $[H]$ are rewritten here as

$$u_{ij} = \delta_{ij} - \frac{k}{2j} \int_{w_{pla,j}} \frac{\boldsymbol{R}}{R} \cdot \hat{\boldsymbol{n}}'_j H_1^{(2)}(kR) dl'_j \tag{3.71}$$

$$h_{ij} = \frac{k\eta d}{2w_{pla,j}} \int_{w_{pla,j}} H_0^{(2)}(kR) dl'_j \tag{3.72}$$

When $R = 0$ or R is very small, the above integrand is singular or semisingular. The following approximation for small arguments of Hankel functions are used to analytically calculate above integrals and fill the matrices $[U]$ and $[H]$.

$$H_1^{(2)}(x) \to \frac{2j}{\pi x} \text{ and } H_0^{(2)}(x) \to \frac{-2j}{\pi} \ln x, \text{ for } x \to 0 \tag{3.73}$$

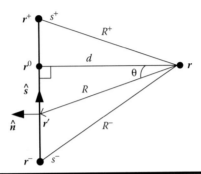

Figure 3.38 **Segment used to analytically calculate the integral singularity.**

A local coordinates are defined for the segment w_j as in Figure 3.38, where r^- and r^+ are the start and end nodes of the segment. r^0 is the projection of the observing point r on the segment. An arc-length coordinate s is defined along the segment with r^0 as its origin and $\hat{s} = (r^+ - r^-)/(|r^+ - r^-|)$ as its positive direction. $\hat{n}$ is the unit outward normal vector of the segment. The length of the segment is $w_j = s^+ - s^-$. Other parameters are defined as

$$\left\{ \begin{array}{l} s^{\pm} \equiv (r^{\pm} - r) \cdot \hat{s} \\[6pt] d \equiv (r^- - r) \cdot \hat{n} \\[6pt] R^{\pm} \equiv \sqrt{s^{\pm 2} + d^2} \\[6pt] R = \sqrt{s^2 + d^2} \\[6pt] \cos(\theta) = \dfrac{d}{R} \end{array} \right. \tag{3.74}$$

In the following, the integrals of Equations 3.71 and 3.72 are divided into the singular part including Equation 3.73 and nonsingular part. The singular parts are analytically calculated by using the following I_1 and I_2, respectively, whereas the nonsingular parts are numerically calculated by using the Gaussian quadrature formula.

The integral I_1 is defined as

$$I_1 \equiv \int_{w_j} \cos\theta \frac{1}{R} dl = \int_{s^-}^{s^+} \frac{d}{d^2 + s^2} ds = \frac{d}{|d|} \left(\tan^{-1} \frac{s^+}{|d|} - \tan^{-1} \frac{s^-}{|d|} \right)$$

$$= \tan^{-1} \frac{s^+}{d} - \tan^{-1} \frac{s^-}{d} \tag{3.75}$$

When $d = 0$, since $\cos\theta = 0$, so $I_1 = 0$.

The matrix filling with I_1 is as follows:

1. For $i = j$, $u_{ij} = 1$.
2. For $R^{\pm} < w_j$ (singular or near singular),

$$u_{ij} = -\frac{k}{2j}\int_{w_j} \cos\theta\left[H_1^{(2)}\left(k|\mathbf{r'}-\mathbf{r}|\right) - \frac{2j}{\pi}\frac{1}{k|\mathbf{r'}-\mathbf{r}|}\right]dl' - \frac{1}{\pi}I_1$$

$$= k\left\{ \int_{w_j} -\frac{1}{2j}\cos\theta\left[H_1^{(2)}\left(k|\mathbf{r'}-\mathbf{r}|\right) - \frac{2j}{\pi}\frac{1}{k|\mathbf{r'}-\mathbf{r}|}\right]dl' - \frac{1}{k\pi}I_1\right\}$$

(3.76)

3. Others

$$u_{ij} = -\frac{k}{2j}\int_{w_j} \frac{\mathbf{R}}{R}\cdot\hat{\mathbf{n}} H_1^{(2)}\left(kR\right)dl' = k\int_{w_j} -\frac{1}{2j}\frac{\mathbf{R}}{R}\cdot\hat{\mathbf{n}} H_1^{(2)}\left(kR\right)dl' \qquad (3.77)$$

The integral I_2 is defined as

$$I_2 \equiv \int_{w_j} \ln\left(kR\right)dl' = \int_{s^-}^{s^+} \ln\left(k\sqrt{d^2+s^2}\right)ds = w_j \ln k + \int_{s^-}^{s^+} \ln(\sqrt{d^2+s^2})ds$$

$$= w_j \ln k + d\left(\tan^{-1}\frac{s^+}{d} - \tan^{-1}\frac{s^-}{d}\right) + s^- - s^+ + s^+ \ln R^+ - s^- \ln R^-$$

$$= w_j\left(\ln k - 1\right) + dI_1 + s^+ \ln R^+ - s^- \ln R^- \qquad (3.78)$$

The matrix filling with I_2 is as follows:

4. For $R^{\pm} < w_j$ (singular or near singular),

$$h_{ij} = \frac{k\eta d}{2}\left\{\frac{1}{w_j}\int_{w_j}\left[H_0^{(2)}\left(kR\right) + \frac{2j}{\pi}\ln(kR)\right]dl' - \frac{2j}{w_j\pi}I_2\right\} \qquad (3.79)$$

5. Others

$$h_{ij} = \frac{k\eta d}{2w_j}\int_{w_j} H_0^{(2)}\left(kR\right)dl' \qquad (3.80)$$

References

1. T. Okoshi, *Planar Circuits for Microwave and Lightwaves.* Springer-Verlag: New York, 1985, pp. 10–42.
2. C. J. Ong, D. Miller, L. Tsang, B. Wu, and C. C. Huang, Application of the Foldy-Lax multiple scattering method to the analysis of vias in ball grid arrays and interior layers of printed circuit boards, *Micro. Opt. Technol. Lett.*, 49(1), 225–231, 2007.
3. Z. Z. Oo, E. X. Liu, E. P. Li, X. C. Wei, Y. J. Zhang, M. Tan, L. W. Li, and R. Vahldieck, A semi-analytical approach for system-level electrical modeling of electronic packages with large number of vias, *IEEE Trans. Adv. Packag.*, 31(2), 267–274, 2008.
4. A. E. Ruehli, Equivalent circuit models for three-dimensional multiconductor systems, *IEEE Trans. Microw. Theory Tech.*, 22(3), 216–221, 1974.
5. E. P. Li, E. X. Liu, L. W. Li, and M. S. Leong, A coupled efficient and systematic full wave time domain macromodeling and circuit simulation for high speed interconnects, *IEEE Trans. Adv. Packag.*, 27(1), 213–223, 2004.
6. A. E. Engin, K. Bharath, M. Swaminathan, M. Cases, B. Mutnury, N. Pham, D. N. D. Araujo, and E. Matoglu, Finite-difference modeling of noise coupling between power/ground planes in multilayered packages and boards, *56th Electr. Compon. and Tech. Conf.*, 1262–1267, 2006.
7. X. C. Wei and E. P. Li, Efficient EMC simulation of enclosures with apertures residing in an electrically large platform using the MM-UTD method, *IEEE Trans. Electromagn. Compat.*, 47(4), 717–722, 2005.
8. X. C. Wei, Y. J. Zhang, and E. P. Li, The hybridization of fast multipole method with asymptotic waveform evaluation for the fast monostatic RCS computation, *IEEE Trans. Antennas Propagat.*, 52(2), 605–607, 2004.
9. J. Y. Fang, Y. Z. Chen, Z. H. Wu, and D. W. Xue, Model of interaction between signal vias and metal planes in electronics packaging, *IEEE 3rd Topical Meeting on Electronic Performance of Electronic Packaging*, Monterey, CA, 1994, pp. 211–214.
10. X. C. Wei, E. P. Li, E. X. Liu, E. K. Chua, Z. Z. Oo, and R. Vahldieck, Emission and susceptibility modeling of finite-size power-ground planes using a hybrid integral equation method, *IEEE Trans. Adv. Packag.*, 31(3), 536–543, 2008.
11. J. Park, H. Kim, Y. Jeong, J. Kim, J. Pak, D. Kam, and J. Kim, Modeling and measurement of simultaneous switching noise coupling through signal via transition, *IEEE Trans. Adv. Packag.*, 29(3), 548–559, 2006.
12. P. A. Kok and D. D. Zutter, Prediction of the excess capacitance of a via-hole through a multilayered board including the effect of connecting microstrips or striplines, *IEEE Trans. Microw. Theory Tech.*, 42(12), 2270–2276, 1994.
13. T. Wang, R. F. Harrington, and J. R. Mautz, Quasi-static analysis of a microstrip via through a hole in a ground plane, *IEEE Trans. Microw. Theory Tech.*, 36(6), 1008–1013, 1998.
14. Y. J. Zhang, J. Fan, G. Selli, M. Cocchini, and D. P. Francesco, Analytical evaluation of via-plate capacitance for multilayer packages or PCBs, *IEEE Trans. Microw. Theory Tech.*, 56(9), 2118–2128, 2008.
15. R. L. Chen, J. Chen, T. H. Hubing, and W. M. Shi, Analytical model for the rectangular power-ground structure including radiation loss, *IEEE Trans. Electromagn. Compat.*, 47(1), 10–16, 2005.
16. HFSS Software, 2017, http://www.ansys.com. 2017.

17. L. Tsang and D. Miller, Coupling of vias in electronic packaging and printed circuit board structures with finite ground plane, *IEEE Trans. Adv. Packag.*, 26(4), 375–384, 2003.

18. C. C. Huang, K. L. Lai, L. Tsang, X. X. Gu, and C. J. Ong, Transmission and scattering on interconnects with via structures, *Micro. Opt. Technol. Lett.*, 46(5), 446–452, 2005.

19. C. Tai, *Dyadic Green's Functions in Electromagnetic Theory*, 2nd ed., IEEE Press: New York, 1994.

20. S. M. Rao, D. R. Wilton, and A. W. Glisson, Electromagnetic scattering by surface of arbitrary shape, *IEEE Trans. Antennas Propagat.*, 30(3), 409–418, 1982.

21. X. C. Wei, E. P. Li, and Y. J. Zhang, Efficient solution to the large scattering and radiation problem using the improved finite element-fast multipole method, *IEEE Trans. Magn.*, 41(5), 1684–1687, 2005.

22. Y. Saad and M. H. Schultz, GMRES: A generalized minimal residual algorithm for solving nonsymmetric linear systems, *SIAM J. Sci. Stat. Comput.*, 7(3), 856–869, 1986.

23. V. Frayssé, L. Giraud, S. Gratton, and J. Langou, A set of GMRES routines for real and complex arithmetics on high performance computers, CERFACS Technical Report TR/PA/03/3, public domain software available at www.cerfacs/algor/Softs, 2003.

24. FEKO Software, 2017, http://www.feko.info/index.html.

25. C. C. Huang, Circuit modeling of power/ground plane structures for printed circuit boards, *Micro. Opt. Technol. Lett.*, 47(1), 97–99, 2005.

26. J. Chen, T. H. Hubing, T. P. Vandoren, and R. E. Dubroff, Power bus isolation using power islands in printed circuit boards, *IEEE Trans. Electromagn. Compat.*, 44(2), 373–380, 2002.

27. I. Ndip, S. Guttowski, and H. Reichl, A Novel interconnected patch-ring (IPR) structure for noise isolation, *IEEE International Symposium on Electromagnetic Compatibility*, Austin, TX, 2009, pp. 328–333.

28. J. Choi, V. Govind, and M. Swaminathan, A Novel electromagnetic bandgap (EBG) structure for mixed-signal system applications, *IEEE Radio and Wireless Conference*, Atlanta, GA, 2004, pp. 243–246.

29. T. K. Wang, C. Y. Hsieh, H. H. Chuang, and T. L. Wu, Design and modeling of a stopband-enhanced EBG structure using ground surface perturbation lattice for power/ground noise suppression, *IEEE Trans. Microw. Theory Tech.*, 57(8), 2047–2054, 2009.

30. Y. Jeong, A. C. W. Liu, L. L. Wai, W. Fan, B. K. Lok, H. Park, and J. Kim, Hybrid analytical modeling method for split power bus in multilayered package, *IEEE Trans. Electromagn. Compat.*, 48(1), 82–94, 2006.

31. G. Feng, Y. J. Zhang, J. L. Drewniak, and L. Zhang, SPICE-compatible cavity and transmission line model for power bus with narrow slots, *2007 International Symposium on Electromagnetic Compatibility*, Hawaii, HA, 2007, pp. 1–5.

32. G. Zou, E. Li, X. Wei, X. Cui, and G. Luo, Modeling of arbitrary power-ground planes with slot by using integral equation and transmission line method, *2009 Asia-Pacific Microwave Conference*, Singapore, 2009.

33. X. C. Wei, E. P. Li, E. X. Liu, and X. Cui, Efficient modeling of re-routed return currents in multilayered power-ground planes by using integral equation, *IEEE Trans. Electromagn. Compat.*, 50(3), 740–743, 2008.

34. X. C. Wei, E. P. Li, E. X. Liu, and R. Vahldieck, Efficient simulation of power distribution network by using integral equation and modal decoupling technology, *IEEE Trans. Microw. Theory Tech.*, 56(10), 2277–2285, 2008.

35. R. F. Harrington, *Field Computation by Moment Methods*. Wiley-IEEE Press: Hoboken, NJ, 1993.

36. D. R. Wilton, S. M. Rao, A. W. Glisson, D. H. Schaubert, O. AI-Bundak, and C. Butler, Potential integrals for uniform and linear source distributions on polygonal and polyhedral domains, *IEEE Trans. Antennas Propagat.*, 32(3), 276–281, 1984.

Chapter 4

Extraction of Via Parameters

Vias are widely used in multilayered printed circuit boards (PCBs) and three-dimensional (3D)/two-and-a-half-dimensional (2.5D) packages to connect horizontal traces on different layers. As a kind of vertical interconnector, via is an essential structure for the circuit integration. However, with the ever-increasing working frequency, the parasitic capacitance and inductance of the via become more obvious than before, which makes the via a major discontinuous structure in PCBs and packages. This discontinuity results in the impedance mismatching along the signal trace and then additional insertion loss, as shown in Chapter 1. For the 2.5D integration with interposers, dense through interposer vias are used. Because the interposer is made up of semiconductor materials, the insertion loss and cross talk of such vertical vias are the major concerns for their electrical design. The long through interposer vias also increase the inductance of the package power distribution network (PDN) and may introduce the power integrity (PI) problem. There is a great demand for the accurate and efficient modeling of vias in PCBs and packages. In this chapter, de-embedded [1] and semianalytical [2] methods are proposed to model the vias used in PCBs and 3D/2.5D packages, respectively.

For the through-hole vias in PCBs, we propose a nonequipotential transmission line model to represent their parasitic circuits. An accurate and efficient de-embedding method is used to extract the parameters inside the models. It accounts for losses and irregular shapes of substrates and vias. Finally, the equivalent circuits of each part of PDN are integrated to perform the system-level signal and PI analysis. The accuracy and efficiency of the proposed method are validated through comparing with the measurement and full-wave analysis.

For the through interposer vias, a semianalytical wideband modeling approach based on cylindrical mode expansion of electromagnetic fields is proposed to analyze the electrical properties of high-density through glass via (TGV) arrays. This method fully captures the scattering and multireflection effects between the vertical cylindrical vias by expanding the electromagnetic waves surrounding the vias as cylindrical wave modes. An equivalent circuit of high-density TGV arrays is derived based on the results of the cylindrical mode expansion method. Meanwhile, the perfect magnetic conductor (PMC) boundary is used in the cylindrical mode method, which greatly simplifies the corresponding equivalent circuit of high-density TGV arrays. The accuracy and efficiency of the proposed field–circuit hybrid method is fully validated by comparing it with 2D quasistatic, 3D full-wave simulations, and measurement results.

4.1 De-Embedding Method for Through-Hole Vias

The PDN inside an electronic package or PCB includes the power–ground planes and the signal traces. The power–ground planes are used to supply the power to integrated circuits with a low-impedance path; the signal traces pass through the power–ground planes to connect integrated circuits to each other as well as to power–ground planes. With increasing operating frequency and power density, the noise coupling inside the PDN becomes very serious, as shown in Chapter 1. Such situations represent a great challenge to commercially available electromagnetic simulators. It is thus desirable to develop alternatives that deliver results fast and with great accuracy [1–6]. The cavity mode theory [1] provides a quick analytic simulation of the power–ground planes. However, it is limited to power–ground planes with regular shapes. Other semi-analytical methods, such as the Foldy-Lax multiple scattering method [5], have been proposed. More accurate modeling of the PDN requires full-wave numerical methods. Although the overall size of the PDN is small enough to apply those full-wave methods, the high aspect ratio of the power–ground planes and the dense and small vias result in a huge mesh. This makes these full-wave methods very expensive in terms of the computing time.

To avoid the computational cost of these full-wave methods, we have proposed a fast integral equation equivalent circuit (IEEC) method [7] in Chapter 3. A two-dimensional (2D) integral equation is used to extract the equivalent circuit of the power–ground planes, and modal decoupling technology is used to integrate signal traces into this equivalent circuit. The IEEC method has been used to model the single plane pair in [7]. This section is an extension of Chapter 3. We extend the IEEC method to model multilayered PDNs. At the same time, an easy de-embedding method is proposed to extract the parasitic equivalent circuit of the multi-through-hole vias.

The through-hole via is an important discontinuous structure in a PDN [8,9]. Usually, quasistatic methods are employed to extract its parasitic circuit. They provide frequency-independent parasitic inductors and capacitors. However, this parasitic circuit is not accurate for high frequencies, where the inductors and capacitors become frequency dependent. In [7], an analytical formula is used to calculate the frequency-dependent values of parasitic inductors and capacitors. Unfortunately, this formula is limited to single vias of cylindrical shapes. This makes the IEEC method unsuitable for more complex PDNs. The de-embedding method proposed in this section provides the frequency-dependent parasitic equivalent circuit and deals with multi-vias of arbitrary shapes and inhomogeneous media.

4.1.1 Vias Modeling

4.1.1.1 Modal Decomposition

Figure 4.1 shows a typical structure of a PDN. It consists of a pair of power–ground planes, a signal trace, and a through-hole via. A via clearance is the void between the via and the power plane, which is also called an "antipad" in some literatures. Different parts of the PDN support different electromagnetic field distributions or modes as shown in Figure 4.2, where the electric field distribution in each part is also plotted.

4.1.1.1.1 Power–Ground Planes

These planes support the parallel-plate mode, which represents the standing wave constrained between the power and ground planes. It is produced by the reflections from the edges of the cavity like pair of power–ground planes. It is the solution of a *two-dimensional* problem defined on the *xy* plane. The equivalent circuit of each pair of power–ground planes can be extracted by using the integral equation method proposed in Chapter 3. In this section, multilayered PDNs are modeled by cascading the equivalent circuits of the power–ground plane pairs.

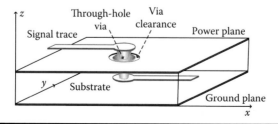

Figure 4.1   A typical PDN.

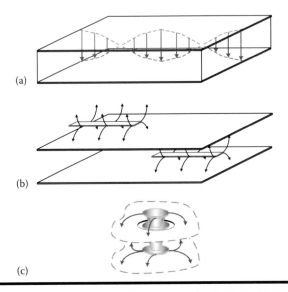

Figure 4.2 Modal decomposition of the electric field in PDN of Figure 4.1: (a) power–ground planes, (b) signal traces, and (c) through-hole via.

4.1.1.1.2 Signal Traces

Signal traces support the transmission mode (transverse electromagnetic mode), which propagates along the traces. After we get their per-unit-length parameters, they are equivalent to the *one-dimensional* transmission lines.

4.1.1.1.3 Through-Hole Vias

Through-hole vias connect the parallel-plate mode and the transmission mode. The vias are much smaller than the working wavelength. They can thus be considered as lumped components, and these are referred as a *zero-dimensional* problem. The parasitic equivalent circuit of the vias will be extracted with the following de-embedding method.

The equivalent circuit for each subproblem is extracted individually. Finally, these equivalent circuits are combined to get the entire solution of the original problem.

4.1.1.2 De-Embedding Method

By using the integral equation procedure explained in Chapter 3, an N-port network is extracted for each pair of power–ground planes. The ports are defined at via clearances of power–ground planes as shown in Figure 4.3. The port current represents the return current of the signal trace as shown in Figure 4.4. It flows from the top plane, passes through the distributed resistance (R), the inductance (L),

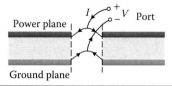

Figure 4.3 **Definitions of port voltage and current for a pair of power–ground planes (cross section).**

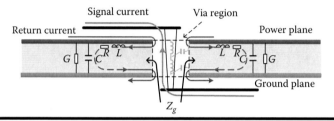

Figure 4.4 **Physical representation of the port current (cross section).**

the capacitance (C), and the conductance (G), and arrives at the bottom plane. The gray region of Figure 4.4 is equivalent to a 2D transmission line (parallel plate waveguide). We define the input impedance seen looking away from the via region as z_g. The return current of Figure 4.4 just follows through z_g. According to the transmission line theory, this z_g is decided by the *LRCG* of the figure and the periphery of the power ground planes. z_g is solved in [7] by assuming there is only fundamental mode (parallel-plate mode) inside the gray region of the figure. The integral domain in [7] does not include the via region (enclosed by the dashed line in Figure 4.4). The via region contains the parasitic inductors and capacitors of the through-hole via, which represent the high-order mode interactions between vias, as well as between vias and power–ground planes.

Because multi-through-hole vias are usually electrically small, their parasitic circuit can be quickly and accurately extracted by using full-wave methods. However, during the calculation, a certain part of the power–ground planes must be included to get the correct parasitic capacitance between the vias and the power–ground planes. These power–ground planes also add the parallel-plate mode effect to the extracted parasitic circuit. We must remove this unwanted parallel-plate mode effect.

Figure 4.5 shows N through-hole vias and a pair of rectangular power–ground planes. We propose a *nonequipotential* multi-transmission lines model to represent the N through-hole vias as shown in Figure 4.6. The vias are divided into small L–C sections, where L means the series inductors, C means the shunt capacitors, and the substrate between the power and ground planes serves as the "reference conductor." Unlike the traditional transmission line model, here z_g of Figure 4.4 is inserted between the reference conductors of adjacent L–C sections. Therefore, the reference

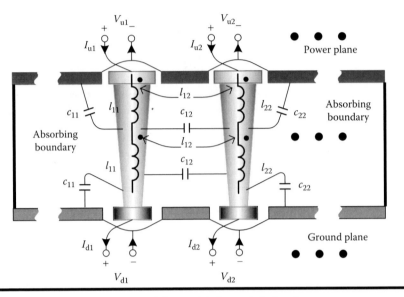

Figure 4.5 The cross section of the multi-through-hole vias.

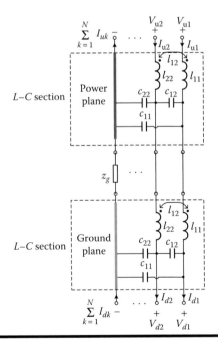

Figure 4.6 Equivalent circuit of multi-through-hole vias.

conductors of each L–C section are nonequipotential. During the extraction, the absorbing boundary condition is placed along the periphery of the power–ground planes. However, due to numerical errors, there is still a small reflection from the periphery. This reflection contributes to z_g. z_g has been solved in [7]. To combine the via with the equivalent circuit of the power–ground planes of [7], we need a parasitic circuit of the multi-through-hole vias with z_g de-embedded.

In Figure 4.6, two identical L–C sections are used. The $ABCD$ (cascade) matrices of the L–C section and z_g are

$$[T]_{\text{vias}} = \begin{bmatrix} U_N + ZY & Z \\ Y & U_N \end{bmatrix} \tag{4.1}$$

and

$$[T]_{\text{PGPs}} = \begin{bmatrix} U_N & Z_g \\ 0 & U_N \end{bmatrix} \tag{4.2}$$

where:

$$[Z] = j\omega \begin{pmatrix} l_{11} & \cdots & l_{1N} \\ \vdots & \ddots & \vdots \\ l_{N1} & \cdots & l_{NN} \end{pmatrix} \tag{4.3}$$

$$[Y] = j\omega \begin{pmatrix} \displaystyle\sum_{k=1}^{N} c_{1k} & -c_{12} \cdots & & -c_{1N} \\ -c_{21} & \displaystyle\sum_{k=1}^{N} c_{2k} \cdots & & -c_{2N} \\ \vdots & \vdots & \ddots & \vdots \\ -c_{N1} & -c_{N2} \cdots & & \displaystyle\sum_{k=1}^{N} c_{Nk} \end{pmatrix} \tag{4.4}$$

$$[Z]_g = \begin{pmatrix} z_g & \cdots & z_g \\ \vdots & \ddots & \vdots \\ z_g & \cdots & z_g \end{pmatrix} \tag{4.5}$$

and $[U]_N$ is the N by N unit matrix. $[Z]$ and $[Y]$ are symmetric matrices. They are unknown matrices to be solved.

The *ABCD* matrix of the entire circuit of Figure 4.6 is defined as

$$[T]_{\text{vias+PGPs}} = \begin{bmatrix} A & B \\ C & D \end{bmatrix} \tag{4.6}$$

For the cascaded networks of Figure 4.6, we get

$$[T]_{\text{vias}} \times [T]_{\text{PGPs}} \times [T]_{\text{vias}} = [T]_{\text{vias+PGPs}} \tag{4.7}$$

$[T]_{\text{vias+PGPs}}$ can be quickly extracted by solving the structure of Figure 4.5 with the finite element method, where only a small area of power–ground planes is involved. z_g is obtained in the same way, where the vias of Figure 4.5 are replaced by a lumped port. Finally, Equations 4.1, 4.2, and 4.6 are substituted into 4.7. By matching the coefficients on both sides of Equation 4.7, we get

$$[Y] = ([U]_N + [D])^{-1} [C] \tag{4.8}$$

$$[Z] = ([B] - [Z]_g)([U]_N + [D])^{-1} \tag{4.9}$$

where [*B*], [*C*], and [*D*] are submatrices defined in Equations 4.6. The *ABCD* matrix of the *L–C* section can be obtained by substituting Equations 4.8 and 4.9 into Equations 4.1. $[T]_{\text{vias}}$ is then converted to the scattering matrix of the *N*-vias. Figure 4.7 shows the extracted C_{11} by using the proposed method and the quasi-static method. The via between the power and ground planes in Figure 4.10 is used. Because a full-wave technology is used to calculate $[T]_{\text{vias+PGPs}}$ and z_g, this proposed de-embedding method presents a frequency-dependent behavior of $[T]_{\text{vias}}$.

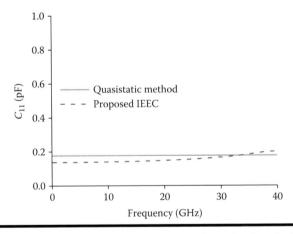

Figure 4.7 **Extracted C_{11} by using the proposed method and the quasistatic method.**

The losses of the substrate and vias can be easily considered by adding the series resistance of the vias and the shunt conductance of the substrate to $[Z]$ and $[Y]$, respectively. In that case, we still can get Equations 4.8 and 4.9.

According to our experiences, when the distance between two vias is larger than three times of the height of the substrate, the mutual capacitors and inductors as shown in Figure 4.5 can be ignored and they can be considered as two individual vias. The extracted parasitic equivalent circuit can be reused for different via layouts. By this way, the proposed IEEC method avoids the repeated computing compared with solving the whole PDN with full-wave methods.

4.1.2 Validation

4.1.2.1 Power–Ground Planes with a Decoupling Capacitor

Figure 4.8 shows a pair of power–ground planes with a surface-mounted decoupling capacitor on the power plane. The decoupling capacitor is used to suppress the parallel-plate mode. The lumped series $\{L, C, R\}$ of the decoupling capacitor are 1.57 nH, 8.14 pF, and 666 mΩ, respectively. Figure 4.9 shows the input impedance of the subminiature A (SMA) port obtained by using the proposed IEEC method and the measurement from [10]. Good agreement can be observed. The proposed method accurately calculates the resonant frequencies of the power–ground planes.

4.1.2.2 Effect of the Through-Hole Via

The through-hole via plays a very important role in the propagation and reflection of the high-speed signal. Figure 4.10 shows the signal trace and the through-hole via under study. The substrate is FR4, with $\varepsilon_r = 4.4$ and a loss tangent of 0.02.

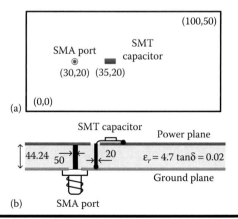

Figure 4.8 A pair of power–ground planes with a decoupling capacitor: (a) top view (unit: mm) and (b) cross section (unit: mils).

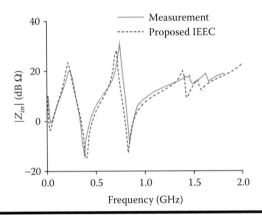

Figure 4.9 Magnitude of the input impedance of SMA port.

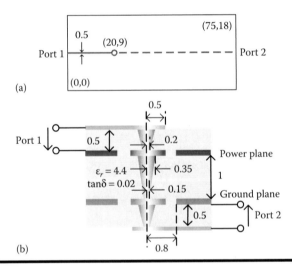

Figure 4.10 A pair of power–ground planes with a signal trace passing through it (unit: mm): (a) top view and (b) cross section.

The thickness of the metal planes and the signal traces is 0.035 mm. The proposed de-embedding method is employed to extract the parasitic circuit of the through-hole via. Two cases are simulated by using the proposed IEEC method: One includes the parasitic circuit of the through-hole via, and the other does not.

The simulated transmission ($|S_{21}|$) and reflection ($|S_{11}|$) characteristics of the signal trace are plotted in Figure 4.11. Results obtained with the commercial full-wave software HFSS (from ANSYS company) are also plotted for comparison. This figure shows that when the proposed de-embedding method is used, the result shows good agreement with the full-wave method. However, if the via discontinuity is not considered, the result greatly deviates from that of the full-wave method,

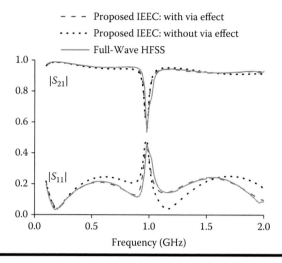

Figure 4.11 Transmission and reflection characteristics of the signal trace in Figure 4.10.

especially at high frequency. Figure 4.11 also reveals that the parasitic circuit of the via can change the wave shape of the reflected signal, but it cannot change the resonant frequency. The resonant frequency is inherently determined by the power–ground planes. In this example, it is related to the TM_{10} mode sustained by the power–ground planes. This shows that the proposed equivalent circuit method clearly accounts for the effects of the different parts in a PDN.

4.1.2.3 Multilayered PDN

In [7], only a single plane pair is simulated. Here, we extend the IEEC method to multilayered PDNs. The example is a PDN with three pairs of power–ground planes as shown in Figure 4.12. There are four copper planes: power plane 1, ground plane 1, power plane 2, and ground plane 2. The substrate is FR4, with $\varepsilon_r = 4.4$ and a loss tangent of 0.02. The thickness of the metal planes and the signal traces is 0.035 mm. All metals are copper. A circular air hole passes through all metal planes and substrates. There are two sets of coupled signal traces. The first set of coupled traces is shown on the top of Figure 4.12a. It includes the coupled microstrip line 1, through-hole vias, and the coupled stripline (CSL). The second set of coupled traces is shown at the bottom of Figure 4.12a. It includes the coupled microstrip line 2, through-hole vias, and the coupled microstrip line 3. One power pin is used to connect power plane 1 and power plane 2. Another ground pin is used to connect ground plane 1 and ground plane 2.

The equivalent circuit of the entire structure is shown in Figure 4.13. Ten ports are defined at the terminals of those coupled microstrip lines and CSLs. The CSLs are split into a set of up CSLs and a set of down CSLs by using the modal decoupling

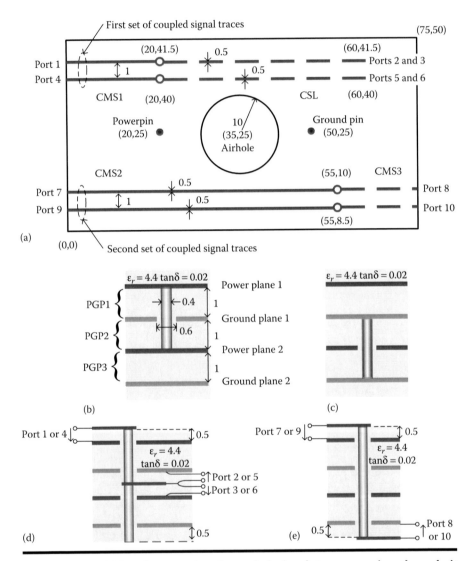

Figure 4.12 PDN and two sets of coupled signal traces passing through it (unit: mm): (a) top view and cross sections of (b) power pin, (c) ground pin, (d) first set of coupled traces, and (e) second set of coupled traces. PGP, power–ground plane pair.

technology [7]. As shown in Figure 4.12d, ports 2 and 5 are defined between the up CSL and the ground plane 1, and ports 3 and 6 are defined between the down CSL and the power plane 2. The equivalent networks of three power–ground plane pairs (which are denoted as PGP1, PGP2, and PGP3 in Figures 4.12 and 4.13) are extracted by using the integral equation. The via equivalent circuits are extracted by using the proposed de-embedding method. The bends that connect the microstrip

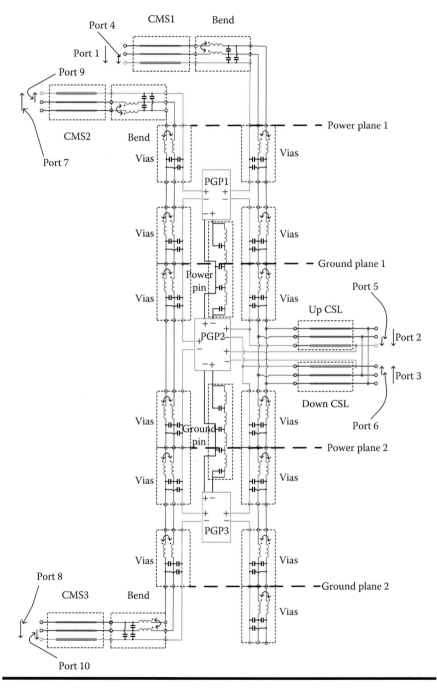

Figure 4.13 Equivalent circuit of the PDN of Figure 4.12. PGP, power–ground plane pair.

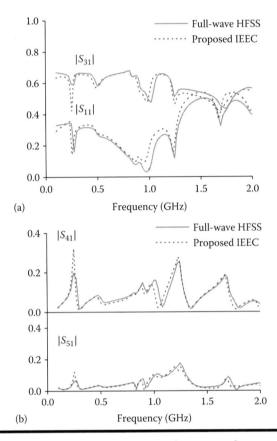

Figure 4.14 **(a) Transmission and reflection; (b) near-end cross talk and far-end cross talk of the first set of coupled traces.**

lines and the vias are modeled as one L–C section. Their equivalent circuits are extracted by using the quasistatic method. All of these equivalent circuits are cascaded together to get the final equivalent circuit of the multilayered PDN.

The transmission ($|S_{31}|$), reflection ($|S_{11}|$), near-end cross talk ($|S_{41}|$), and far-end cross talk ($|S_{51}|$) of the first set of coupled traces are plotted in Figure 4.14. The transmission ($|S_{87}|$), reflection ($|S_{77}|$), near-end cross talk ($|S_{97}|$), and far-end cross talk ($|S_{107}|$) of the second set of coupled traces are plotted in Figure 4.15. Figure 4.16 shows the cross talk ($|S_{81}|$) between the first and second sets of coupled traces. The results from the full-wave HFSS simulation are also plotted for validation. For this complex multilayered PDN, the proposed IEEC method shows good agreement with the full-wave solver.

For the high frequency, there is a little difference between the IEEC method and the full-wave simulation. This is because that the radiation from the microstrip lines and the periphery of the power–ground planes become strong at the high frequency. This radiation is neglected in the current proposed method. Due to this radiation,

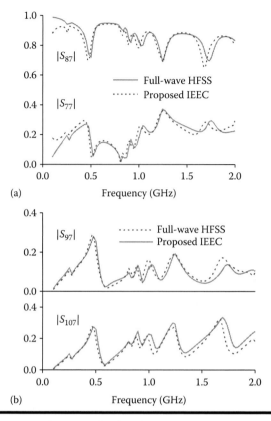

(a)

(b)

Figure 4.15   (a) Transmission and reflection; (b) near-end cross talk and far-end cross talk of the second set of coupled traces.

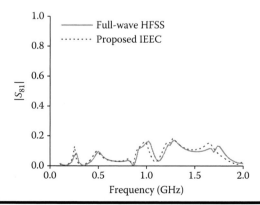

Figure 4.16   Cross talk between the first and the second set of coupled traces.

Table 4.1 Computing Times for the IEEC Method and Full-Wave Simulator HFSS

Example	IEEC	HFSS
B	6 minutes	55 minutes
C	10 minutes	120 minutes

the peak in the HFSS result of Figure 4.14b is also smaller than that in the IEEC result. Radiative models of microstrip lines and power–ground planes can be used to reduce this difference.

4.1.2.4 Comparison of Computing Time

Table 4.1 lists the computing times of the proposed IEEC method and the full-wave simulator HFSS, for example, *B* and *C*. The computing time of the IEEC method includes the time needed to extract the equivalent circuit of each part and to combine them together. It appears that the IEEC method greatly reduces the computing time compared with HFSS. HFSS is based on the 3D finite element method. It needs a dense volume mesh to accommodate the large aspect ratio of the power–ground planes and the tiny structures such as the through-holes and traces. This results in a large number of unknowns. In the IEEC method, we decompose the complex 3D problem into simple 2D, one-dimensional, and zero-dimensional subproblems, so that the computing efficiency is greatly increased.

4.1.3 Conclusion

In this section, we proposed an efficient IEEC method to simulate the electrical performance of PDNs in multilayered PCBs. A de-embedding method is used to extract the parasitic circuit of through-hole vias. The major advantage of the proposed method resides in the decomposition of the complex 3D problem into several simple 2D, one-dimensional, and zero-dimensional subproblems, so that the computing time is greatly reduced while a good accuracy is presented. With the proposed method, the scattering parameters of the PDN can be calculated, where the ports are defined at the critical parts with potential interference problems. Based on the obtained scattering parameters, the model order reduction can be used to reduce the original equivalent circuit to a simple circuit. This will further increase the computing efficiency.

4.2 Cylindrical Mode Expansion Method for TGVs

3D hybrid integration technology that integrates heterogeneous chips, such as processor, memory, digital, analogy, and integrated passives devices, is promising to provide high-speed and high-performance electronic devices [11]. Silicon interposer is widely employed in 3D and 2.5D integration. However, it suffers some electromagnetic compatibility (EMC) problems, such as larger insertion loss and noise coupling [12,13]. Some methodologies have been presented to suppress such substrate loss and solve those EMC issues [14–17] with a more complex process and higher fabrication cost.

In consideration of the disadvantages of silicon-based 3D integration, recently, some glass companies, including Corning, have proposed large, thin, and high-performance glass wafers for interposer applications [12,18]. Glass interposers with TGVs provide a wide range of better properties compared to silicon interposers, such as chemical and mechanical durability, cost reduction, reduced process complexity, and better electromagnetic properties. The high electrical resistivity of glass interposer avoids the use of dielectric layer between copper via and substrate, which not only provides cost reduction and reduced process complexity but also improves the electromagnetic properties of the system compared to the silicon interposer [13,19,20]. All these excellent advantages make glass interposer a promising replacement of silicon interposer for the 3D/2.5D integration.

To improve the performance of the system, it is of great importance to analyze the electrical properties of through vias in glass interposers. So far, some methodologies have been proposed to analyze the electrical characteristics of through-via interposers. The equivalent circuits for limited numbers of through-silicon vias (TSVs) were proposed in [21–25], where different closed-form expressions of electrical parameters were presented. These closed-form expressions resulted from different quasistatic assumptions of the TSVs. Kim et al. [21] undervalued the inductance of the through-via. However, the methodologies proposed in [24] and [25] overestimated the coupling capacitance. Papers [22,23] based on two-conductor transmission lines theory were only effective for the configuration of two TSVs, which will result in some error when applied to through-via arrays with high density. The modeling approach based on multiconductor transmission line theory proposed in [26–28] could capture the electrical properties of multiple conductors. Fourier–Bessel expansion approach presented in [29] was applied to solve ground-signal (GS) configuration. Cylindrical modal basis functions proposed in [30–33] captured the multiple scattering characteristics among vias, where the additional top and bottom ground planes are required to be added on the interposer.

The modeling of TSV- and TGV-based interposers is different due to the different geometrical structures of TSVs and TGVs. Although there had been lots of results published about TSV electrical modeling, the electrical properties of high-density TGV arrays are not fully investigated.

In this section, a field–circuit hybrid method is presented for the electromagnetic characteristic analysis of high-density TGV arrays. The method is based on cylindrical mode expansion [34]. The value of the lumped $RLGC$ parameters can be calculated by using this method. Meanwhile, the PMC boundary is used in the proposed method to simplify the resulting equivalent circuits.

4.2.1 Field–Circuit Hybrid Method

The field–circuit hybrid method for modeling high-density TGV arrays is proposed in this section. In the following, TGV arrays containing $N+1$ randomly distributed conductors are considered to demonstrate the proposed method. One of the conductors is selected as the reference conductor (denoted as the #0 conductor) for return current and the selection is not unique. The other N TGVs are considered as signal conductors.

The length of the TGVs is tens of micrometers, which is short enough compared to the wavelength in gigahertz band in the glass substrate. Therefore, the TGV arrays can be modeled with a lumped equivalent circuit model as shown in Figure 4.17. The electrical parameters $RLGC$ in the equivalent circuit model of $N+1$ TGVs are per-unit-length resistance, inductance, capacitance, and conductance, respectively.

4.2.1.1 Internal Self-Impedance of TGV

The rigorous closed-form expression for the resistance of TGVs consists of *dc* and *ac* resistances. The *dc* resistance is obtained from Ohm's law, which can be depicted as

$$R_{dc} = \frac{\rho}{\pi r^2} \tag{4.10}$$

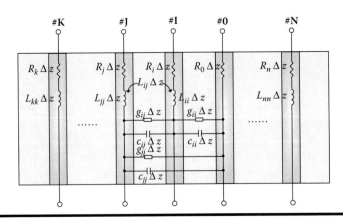

Figure 4.17 Equivalent circuit model of TGV arrays that contain $N+1$ conductors.

As the frequency increases, *ac* resistance must be considered. The analytical expression is given as

$$R_{ac} = \frac{\rho}{\pi\delta(2r - \delta)} \tag{4.11}$$

where:
$\delta = \sqrt{1/\pi\sigma f \mu_0}$
δ denotes the skin depth
ρ is the resistivity of copper
R is the radius of TGVs
F is the frequency
μ_0 denotes the permeability in vacuum.

The *internal self-inductance L_i* of TGV can be expressed as

$$L_i = \frac{R_{ac}}{2\pi f} \tag{4.12}$$

The internal self-impedances of an isolated TGV and TGV in an array are the same. However, their *external self-impedances* are different. The external self-impedance of TGV will be changed by its surrounding TGVs. The external self-inductance together with other parasitic parameters will be calculated in Section 4.2.1.2 by using the cylindrical mode expansion method.

4.2.1.2 Cylindrical Mode Expansion Method

Because the total inductance is the summation of internal inductance and external inductance, a detailed description of external self-inductance and mutual inductance by the cylindrical mode expansion method will be given in this section. After that, the capacitance and conductance matrices can be calculated by the inductance matrix.

The multiple scattering of electromagnetic waves in the interposer between TGV arrays is complex. It is difficult for the analytical expressions mentioned in the available publications to accurately model this behavior. The proposed method that is based on the cylindrical mode expansion fully considers the scattering and multi-reflection effects between the vertical cylindrical vias by expanding the electromagnetic waves surrounding the vias as cylindrical modes. After that, an accurate equivalent circuit based on electrical parameters that are calculated by the proposed cylindrical mode expansion method can be obtained as shown in Figure 4.17.

The per-unit-length inductance matrix L relates the total magnetic flux penetrating the loop between the ith TGV and the reference TGV, per unit of via length, to all the via currents I producing it as

$$\Psi = LI \tag{4.13}$$

and its matrix form can be expressed as

$$
\begin{bmatrix} \psi_1 \\ \psi_2 \\ \vdots \\ \psi_n \end{bmatrix}
=
\begin{bmatrix}
l_{11} & l_{12} & \cdots & l_{1n} \\
l_{12} & l_{22} & \cdots & l_{2n} \\
\vdots & \vdots & \ddots & \vdots \\
l_{1n} & l_{2n} & \cdots & l_{nn}
\end{bmatrix}
\begin{bmatrix} I_1 \\ I_2 \\ \vdots \\ I_n \end{bmatrix}
\tag{4.14}
$$

The components l_{ii} are the *external self-inductances* of the ith TGV, and the components l_{ij} are the mutual inductances between the ith TGV and the jth TGV. They are defined as

$$l_{ii} = \frac{\psi_i}{I_i}\bigg|_{I_1=\cdots=I_{i-1}=I_{i+1}=I_n=0}$$

$$l_{ij} = \frac{\psi_i}{I_j}\bigg|_{I_1=\cdots=I_{j-1}=I_{j+1}=I_n=0} \tag{4.15}$$

We can compute the L matrix by applying the current in *one loop* (i.e., signal current on one TGV, and its return current on the reference TGV) and setting the currents on all the other TGVs to zero, then determining the total magnetic flux penetrating this loop of per unit length. It should be noted that the values of l_{ii} and l_{ij} are dependent with the entire layout of the TGV array, which cannot be accurately calculated by using available closed-form expressions based on two-conductor transmission lines.

Considering that an *ac* current I distributes over the ith TGV and currents along other TGVs are set to zero. The excited transverse magnetic wave by the excitation current can be expressed by the longitudinal field E_z in the through-via local cylindrical coordinate as shown in Figure 4.18.

Under the cylindrical coordinates (ρ, φ, z), the electric field E_z can be expressed as

$$E_z = \sum_{n=-\infty}^{\infty} \left[a_n J_n(k\rho) + b_n H_n^{(2)}(k\rho) \right] e^{jn\varphi} \tag{4.16}$$

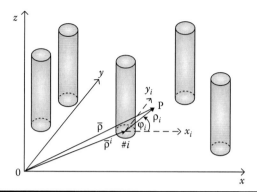

Figure 4.18   Schematic of the randomly distributed TGV arrays.

where:

a_n and b_n are the expansion coefficients of the incoming wave and the outgoing wave, respectively

k is the wavenumber with the value of $\omega\sqrt{\mu\varepsilon}$

ω is the angular frequency

μ and ε are the permeability and permittivity of the glass interposer, respectively

$J_n(k\rho)$ and $H_n^{(2)}(k\rho)$ are the Bessel function with the order of n and the second kind Hankel function with the order of n, respectively

The tangent component of the cylindrical mode is governed by the following equation:

$$H_t = -\frac{1}{j\omega\mu}\nabla\times E = -\frac{1}{j\omega\mu}\left(\frac{1}{\rho}\frac{\partial E_z}{\partial\varphi}\hat{\rho} - \frac{\partial E_z}{\partial\rho}\hat{\varphi}\right) \tag{4.17}$$

By substituting Equation 4.16 into 4.17, the magnetic fields are obtained as

$$H_\rho = -\frac{1}{\omega\mu\rho}\sum_{n=-N}^{N} n\left[a_n J_n(k\rho) + b_n H_n^{(2)}(k\rho)\right]e^{jn\varphi} \tag{4.18}$$

$$H_\varphi = \frac{k}{j\omega\mu}\sum_{n=-N}^{N}\left[a_n J_n'(k\rho) + b_n H_n^{'(2)}(k\rho)\right]e^{jn\varphi} \tag{4.19}$$

The excited electromagnetic waves in the glass interposer form multiple scattering among the high-density TGVs. As a result, for a set of randomly distributed TGVs with different radii, the total electromagnetic fields at the observation point p can be expressed as the summation of the outgoing waves from all the scattering vias as

$$\phi = \sum_{i=1}^{N+1}\sum_{n=-N_i}^{n=N_i} f_{in} H_n^{(2)}(k\rho_i)e^{jn\varphi_i} \tag{4.20}$$

where $2N_i+1$ is the truncated order of Hankel functions. ρ_i and φ_i are defined in the local cylindrical coordinate of the ith via as shown in Figure 4.18. f_{in} is the unknown coefficient of each wave mode and can be derived from the following equation:

$$\boldsymbol{f}_i = \overline{T}_i(\boldsymbol{a}_i + \sum_{j=1;j\neq i}^{N+1} \overline{\boldsymbol{\alpha}}_{ij}\boldsymbol{f}_j) \qquad (4.21)$$

where:
   $\overline{T}_i$ is the reflection coefficient of the ith via
   $\boldsymbol{a}_i$ denotes the incident wave coefficients of the excited sources in terms of Bessel functions

The incident wave of the ith via caused by the scattered wave from the jth via can be represented by the translation matrix $\overline{\boldsymbol{\alpha}}_{ij}$. By using addition theorem of cylindrical harmonics, $\overline{\boldsymbol{\alpha}}_{ij}$ can be obtained as

$$\overline{\alpha}_{ij}(m,n) = H^{(2)}_{m-n}\left(k\rho_{ji}\right)e^{-j(m-n)\varphi_{ji}} \qquad (4.22)$$

where $m = \left[-M_i, M_i\right]$, $n = \left[-M_j, M_j\right]$. m and n are the number of expansion coefficients that satisfy the sufficient accuracy. The symbols ρ_{ji} and φ_{ji} are defined as $\rho_{ji} = \left|\overline{\rho}^j - \overline{\rho}^i\right|$ and $\varphi_{ji} = \arg(\overline{\rho}^j - \overline{\rho}^i)$; $\overline{\rho}^j$ and $\overline{\rho}^i$ present the positions of the jth and ith TGVs, respectively.

$\overline{T}_i$ can be obtained from the boundary condition of TGVs. The boundary conditions of TGVs include the perfect electric conductor and PMC. According to the definition of l_{ii} and l_{ij} in Equation 4.15, the current on nonexcitation TGVs is set to zero so that those TGVs can be considered as PMC in the cylindrical mode expansion method, where the tangential component of the magnetic field is set to zero. The effect of the entire layout of the TGV arrays on l_{ii} and l_{ij} can be accurately considered by those PMC vias. Applying the PMC boundary condition at the interface between TGV and the interposer, the reflection matrix can be expressed as

$$\left[\overline{T}_i\right]_{nn} = \frac{b_n}{a_n} = -\frac{J'_n(kr)}{H'^{(2)}_n(kr)} \qquad (4.23)$$

The usage of PMC boundaries in the derivation of l_{ii} and l_{ij} not only captures the physical mechanism but also simplifies the resulted equivalent circuit. Heat transferring TGVs and electromagnetic-shielding TGVs are used in practical glass interposer design for heat dissipation and noise reduction, respectively. These TGVs can be floating and open circuited; therefore, no ports are defined for them in the equivalent circuit. This greatly reduces the number of ports in the final equivalent circuit and speed up the subsequent circuit simulation. At the same time, the field perturbation caused by the adjacent heat transferring

TGVs and shielding TGVs can be accurately included in the cylindrical mode expansion method by using the PMC vias.

The total magnetic flux of per-unit-length through the surface between the ith TGV and the reference TGV calculated by the cylindrical mode expansion method can be expressed as

$$\psi = u \int_c \hat{H}_t \cdot \hat{a}_n \, dl = u \int_c \sum_{i=1}^{N+1} (H_{\rho i} \hat{\rho}_i + H_{\varphi i} \hat{\varphi}_i) \cdot \hat{a}_n \, dl \qquad (4.24)$$

where:

$H_{\rho i}$ and $H_{\varphi i}$ are the ρ and φ components of the scattering magnetic fields by the ith through via, respectively, which can be calculated by Equation 4.20

$\hat{a}_n$ is the normal vector of the integral line c

By substituting Equation 4.24 into 4.15, considering the internal self-inductance and the contribution from both the signal current on one TGV and its return current on the reference TGV, the per-unit-length inductance matrix can be obtained.

The parallel per-unit-length capacitance and conductance matrix of the TGV arrays in the glass interposer can be calculated by the following relation [35]:

$$[L][C] = \mu\varepsilon I_N \qquad (4.25)$$

$$[G] = \frac{\sigma}{\varepsilon}[C] \qquad (4.26)$$

where:

I_N is the unit matrix

σ and ε are the conductivity and permittivity of the glass interposer, respectively

4.2.2 Simulation Results

The electrical properties of the TGVs with the existence of other TGVs in the vicinity can be different from that of the isolated TGV. The proposed field–circuit hybrid method fully considers this effect in the final equivalent circuit model. In this section, the proposed field–circuit hybrid method will be utilized to model TGV arrays and validate the accuracy and efficiency by comparing the results with available closed-form expressions based on two-conductor transmission lines and full-wave simulation.

4.2.2.1 Per-Unit-Length Inductance

To validate the accuracy of the proposed field–circuit hybrid method based on the cylindrical mode expansion for calculating per-unit-length electrical parameters, a case of nine coupled TGVs is given as shown in Figure 4.19. The central TGV is set as the reference conductor for the return current and other TGVs are signal conductors.

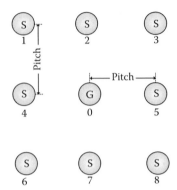

Figure 4.19 Top view of nine coupled TGVs where the middle TSV is set as the reference conductor.

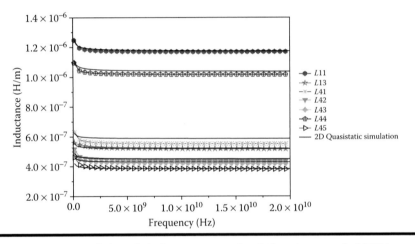

Figure 4.20 Per-unit-length inductance matrix of the nine coupled TGV array. Solid lines denote the quasistatic simulations results and marks denote the proposed method results.

The radius of TGVs is 15 μm and the center-to-center pitch is 200 μm. The relative permittivity and loss tangent of the glass interposer are 6.7 and 0.006, respectively.

The 2D quasistatic simulation is applied to calculate the per-unit-length inductance matrix and validate the efficiency and accuracy of the proposed cylindrical mode expansion method up to 20 GHz. The results are shown in Figure 4.20. Very good consistency can be obtained between the value extracted from the 2D simulator and the cylindrical mode expansion method.

4.2.2.2 Signal–Ground–Signal TGVs

The signal–ground–signal (SGS) configuration which is the fundamental structure for signal transmission is chosen for the verification of the presented method as

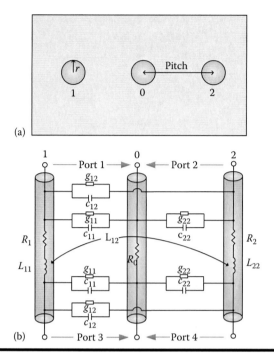

Figure 4.21 **(a) Top view of the SGS-type TGV structure and (b) the derived equivalent circuit model by the proposed method.**

shown in Figure 4.21a. The radius of TGVs is 10 μm and the center-to-center pitch between the signal TGVs and the ground TGV is 100 μm. The height of TGVs is 100 μm. A π-type equivalent circuit is established in Figure 4.21b. The electrical parameters inductance, capacitance, and conductance are calculated by the proposed cylindrical mode expansion method, and the values of the conductance and capacitance in the established equivalent circuit are half of the results calculated in Section 4.2.1. Resistance is calculated by analytical formulas found in Section 4.2.1.

The transmission coefficient, reflection coefficient, and cross talk obtained from the proposed method are compared with 3D full-wave simulation and available closed-form expressions in Figure 4.22. Very good correlation is obtained between the proposed method and 3D full-wave simulator. However, there exist some distinctions between the available closed-form expressions in the publications and the full-wave method at high frequencies. It is because the available closed-form expressions do not consider the variation of capacitance and conductance of the original GS pair TGVs in vicinity of other TGVs. This can lead to the limits of accuracy and erroneous results. The method proposed in this section fully considers the multi-scattering among the TGVs in the array; hence, the presented method can fully capture the signal transmission and noise coupling mechanism.

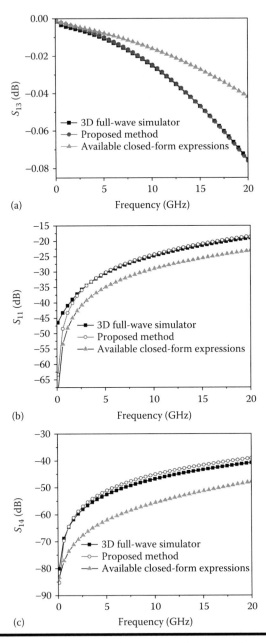

(a)

(b)

(c)

Figure 4.22 Comparison of electrical properties between the proposed method, 3D full-wave simulator, and the results obtained from the available closed-form expressions.

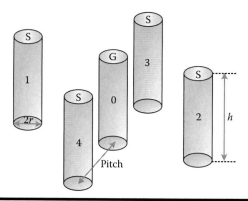

Figure 4.23 Full view of the cruciform structure with four signal TGVs.

4.2.2.3 Multiple Signal TGV Array

In this section, a case of cruciform structure configuration of TGV arrays with four signal TVGs is given to validate the accuracy of the proposed field–circuit hybrid method. The central #0 TGV is set as the reference conductor for the return current and other TGVs are signal conductors as shown in Figure 4.23.

The radius and height of the TGVs are 15 and 200 μm, respectively. The center-to-center pitch between the signal TGVs and the ground TGV is set to 200 μm. Then, the electrical characteristics of the system captured by the equivalent circuit model based on the proposed method are compared with that from 3D full-wave simulator. It can be seen in Figure 4.24 that the simulation results based on the proposed field–circuit hybrid method have good correlation with that obtained from the 3D full-wave simulator.

4.2.2.4 TGV Arrays with Floating TGVs

Heat transferring TGVs and shielding TGVs are widely used in interposer application. Conversional closed-form methods do not consider the variation of the electromagnetic fields in vicinity of those additional TGVs; thus, they cannot give an accurate equivalent circuit model. The proposed field–circuit hybrid method can fully consider the electromagnetic wave scattering with additional TGVs. A case of SGS configuration with additional heat transferring TGVs and shielding TGVs inserted is shown in Figure 4.25, where ports 1–4 are also defined. The dark TGVs represent the ground and signal TGVs, whereas the other TGVs represent the additional TGVs that are floating.

These open-circuited floating TGVs are modeled as PMC vias, which not only provide the accurate lumped *RLGC* parameters but also greatly simplify the corresponding equivalent circuit of high-density TGV arrays. The variation of the electromagnetic waves caused by floating TGVs in the interposer can be accurately modeled by PMC boundaries. In this case, the proposed method fully captures the practical physical mechanism and considers this influence in the equivalent circuit.

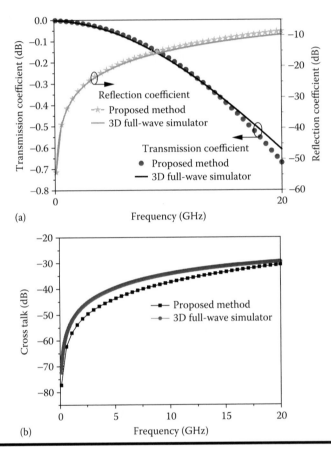

(a)

(b)

Figure 4.24 Validation of the TGV array: (a) insertion loss and reflection coefficient validation and (b) far-end cross talk validation.

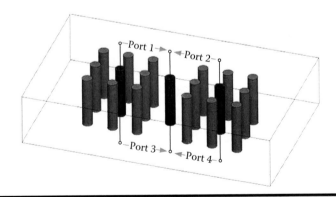

Figure 4.25 TGV array structure with floating heat transferring TGVs and shielding TGVs.

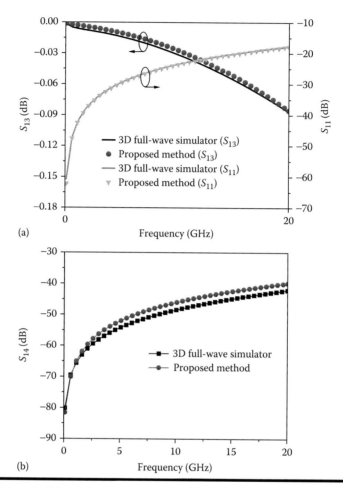

Figure 4.26 Validate the proposed method by comparing the results of the proposed method with that of the 3D full-wave simulator: (a) insertion loss and reflection coefficient validation and (b) far-end cross talk validation.

As depicted in Figure 4.26, the results obtained from the proposed method agree well with that of the 3D full-wave simulator. All the simulations are conducted on a computer with 3.3 GHz Intel Xeon CPU and 8GB of RAM memory. The simulation time required by the proposed cylindrical mode expansion method is six seconds, in contrast to 10 minutes by 3D full-wave simulator on the same computer. These verify the efficiency of the proposed field–circuit hybrid method.

4.2.2.5 TGV with RDL

The redistribution layer (RDL) that provides horizontal interconnects between different vertically stacked layers to redistribute the signals is needed in practical 3D

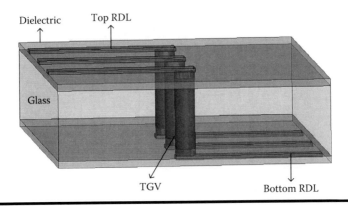

Figure 4.27 SGS configuration of TGV arrays with RDL interconnects.

packaging. In this section, the SGS type of TGVs together with RDL interconnects will be modeled as shown in Figure 4.27.

The radius of TGVs is 15 μm and the center-to-center pitch between the signal TGVs and the ground TGV is 200 μm. A thickness of 10 μm dielectric layer is inserted between the RDL and the 100 μm height of glass substrate. The width of the RDL interconnects are 40 μm with the length of 215 μm. By establishing the equivalent circuit model of TGVs using the proposed field–circuit hybrid method and the RDL using analytical expressions, the electrical characteristics can be obtained and compared with the 3D full-wave simulator.

As can be seen in Figure 4.28, in the whole frequency band, the electrical parameters of the transmission coefficient, reflection coefficient, and cross talk obtained from the proposed method and full-wave simulation are in good agreement.

4.2.3 Experiment Validation

A nine-stacked silicon interposer with TSV arrays published in [36], as shown in Figure 4.29, is used here for the validation. Micro-bumps are used to assemble two different vertical dies. Because the proposed field–circuit hybrid method is a general method, it can be used for both TGV arrays and TSV arrays.

The 3D view of the nine-stacked interposer and the top view of TSVs can be seen in Figure 4.30. The radius r of TSV is 30 μm, the thickness of the insulator is 20 μm, and the height of interposer is 94 μm. Among all of TVSs, three TSVs (#0, #1, and #2) are selected to define the ports as shown in Figure 4.30. The center-to-center pitch is 630 μm for TSV 0 and TSV 2, and 500 μm for TSV 0 and TSV 2.

In Figure 4.31, S_{13} represents the transmission coefficient and S_{12} represents the near-end cross talk. The measurement frequency is from *dc* to 8 GHz in [36]. Good agreement can be seen between the measurement data and the proposed method in both insertion loss and cross talk coefficient.

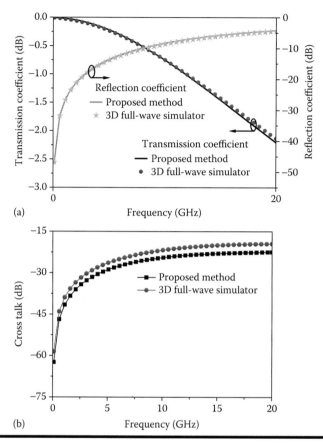

(a)

(b)

Figure 4.28 Validation of TGV arrays with RDL interconnects: (a) insertion loss and reflection coefficient validation and (b) near-end cross talk validation.

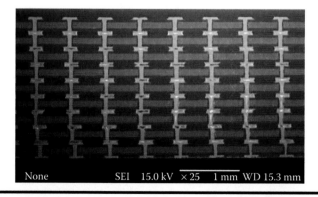

Figure 4.29 Scanning electron microscope photo of the nine-stacked silicon interposer. (From C. D. Wang et al., *IEEE Trans. Comp. Packag. Manuf. Technol.*, 3(10), 1744–1753, 2013.)

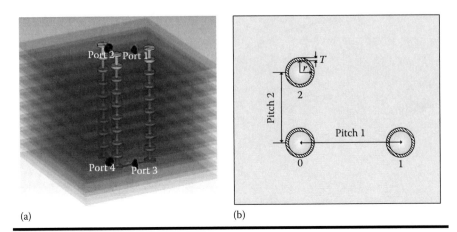

(a)   (b)

Figure 4.30 (a) 3D view of the nine-stacked interposer and (b) the structure diagram of TSVs.

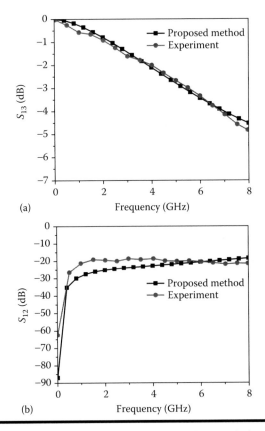

(a)

(b)

Figure 4.31 Comparison between measurement data and simulation results of the proposed method for (a) insertion loss and (b) near-end cross talk.

4.2.4 Conclusion

An accurate field–circuit hybrid method for modeling high-density TGV arrays is presented in this section. The method is based on the cylindrical mode expansion which considers scattering and multi-reflection effects between the vertical cylindrical vias. Based on the proposed method, accurate electrical parameters can be calculated and the resulting equivalent circuit can be greatly simplified. As the method fully captures the physical mechanism of the wave interaction between multiple vias, the calculated transmission coefficient and cross talk agree well with that of 3D full-wave simulation. The measurement data of a nine-stacked interposer also agree well with that of the proposed method.

References

1. X. C. Wei and E. P. Li, Integral-equation equivalent-circuit method for modeling of noise coupling in multilayered power distribution networks, *IEEE Trans. Microw. Theory Tech.*, 58(3), 559–565, 2010.
2. J. Li, X. C. Wei, and E. P. Li, Accurate field-circuit hybrid modeling of high-density through glass via arrays by using perfect magnetic conductors and cylindrical mode expansion, *IEEE Trans. Comp. Packag. Manuf. Technol.*, 6(1), 100–108, 2015.
3. T. Okoshi, *Planar Circuits for Microwave and Lightwaves*. Springer-Verlag: New York, 1985, pp. 10–42.
4. A. E. Engin, K. Bharath, M. Swaminathan et al., Finite-difference modeling of noise coupling between power/ground planes in multilayered packages and boards, in *56th Electronic Components and Technology Conference*, San Diego, CA, 2006, pp. 1262–1267.
5. C. J. Ong, D. Miller, L. Tsang et al., Application of the Foldy-Lax multiple scattering method to the analysis of vias in ball grid arrays and interior layers of printed circuit boards, *Micro. Opt. Technol. Lett.*, 49(1), 225–231, 2007.
6. Z. Z. Oo, X. C. Wei, E. X. Liu et al., Efficient analysis for multilayer power ground planes with multiple vias and signal traces in an advanced electronic package, in *Electrical Performance of Electronic Packaging.*, San Jose, CA, 2008, pp. 95–98.
7. X. C. Wei, E. P. Li, E. X. Liu et al., Efficient simulation of power distribution network by using integral equation and modal decoupling technology, *IEEE Trans. Microw. Theory Technol.*, 56(10), 2277–2285, 2008.
8. J. Park, H. Kim, Y. Jeong et al., Modeling and measurement of simultaneous switching noise coupling through signal via transition, *IEEE Trans. Adv. Packag.*, 29(3), 548–559, 2006.
9. P. A. Kok and D. D. Zutter, Prediction of the excess capacitance of a via-hole through a multilayered board including the effect of connecting microstrips or striplines, *IEEE Trans. Microw. Theory Technol.*, 42(12), 2270–2276, 1994.
10. J. Fan, J. L. Drewniak, J. L. Knighten et al., Quantifying SMT decoupling capacitor placement in DC power-bus design for multilayer PCBs, *IEEE Trans. Electromagn. Compat.*, 43(4), 588–599, 2001.
11. S. J. Bleiker, A. C. Fischer, G. Stemme et al., High-aspect-ratio through silicon vias for high-frequency application fabricated by magnetic assembly of gold-coated nickel wires, *IEEE Trans. Comp. Packag. Manuf. Technol.*, 5(1), 21–27, 2015.

12. M. Lee, J. Cho, J. Kim et al., Noise coupling of through-via in silicon and glass interposer, in *Electronic Components and Technology Conference*, Las Vegas, NV, 2013, pp. 1806–1810.

13. V. Sukumaran, T. Bandyopadhyay, V. Sundaram et al. Low-cost thin glass interposers as a superior alternative to silicon and organic interposers for packaging of 3-D ICs, *IEEE Trans. Comp. Packag. Manuf. Technol.*, 2(9), 1426–1433, 2012.

14. J. Cho, E. Song, K. Yoon et al., Modeling and analysis of through-silicon via (TSV) noise coupling and suppression using a guard ring, *IEEE Trans. Comp. Packag. Manuf. Technol.*, 1(2), 220–233, 2011.

15. J. J. Tang, X. Chen, G. W. Xu et al., A novel wafer-level metal/BCB interconnection between both sides of wafer using TSV and its microwave performance, in *Electronic Components and Technology Conference*, San Diego, CA, 2012, pp. 2121–2128.

16. C. Huang, Q. W. Chen, D. Wu et al., Implementation of air-gap through-silicon-vias (TSVs) using sacrificial technology, *IEEE Trans. Comp. Packag. Manuf. Technol.*, 3(8), 1430–1438, 2013.

17. Q. W. Chen, C. Huang, Z. M. Tan et al., Low capacitance through-silicon-vias with uniform benzocyclobutene insulation layers, *IEEE Trans. Comp. Packag. Manuf. Technol.*, 3(5), 724–731, 2013.

18. C. Kim and Y. K. Yoon, High frequency characterization and analytical modeling of through glass via (TGV) for 3D thin-film interposer and MEMS packaging, in *Electronic Components and Technology Conference*, San Diego, CA, 2013, pp. 1385–1391.

19. J. Keech, S. Chaparala, A. Shorey et al., Fabrication of 3D-IC interposers, in *Electronic Components and Technology Conference*, San Diego, CA, 2013, pp. 1829–1833.

20. B. K. Wang, Y. A. Chen, A. Shorey et al., Thin glass substrates development and integration for through glass vias (TGV) with Cu interconnect, in *Electronics Packaging Technology Conference*, Singapore, 2012, pp. 351–354.

21. J. Kim, J. S. Pak, J. Cho et al., High-frequency scalable electrical model and analysis of a through silicon via (TSV), *IEEE Trans. Comp. Packag. Manuf. Technol.*, 1(2), 181–195, 2011.

22. I. Ndip, K. Zoschke, K. Lobbicke et al., Analytical, numerical-, and measurement-based methods for extracting the electrical parameters of through silicon vias (TSVs), *IEEE Trans. Comp. Packag. Manuf. Technol.*, 4(3), 504–515, 2014.

23. Z. Xu and J. Lu, Through-silicon-via fabrication technologies, passives extraction, and electrical modeling for 3-D integration/packaging, *IEEE Trans. Semicond. Manuf.*, 26(1), 23–34, 2013.

24. J. Kim, J. Cho, J. Kim et al., High-frequency scalable modeling and analysis of a differential signal through-silicon via, *IEEE Trans. Comp. Packag. Manuf. Technol.*, 4(4), 697–707, 2014.

25. Q. H. Li and M. Miao, Study of ground-signal-ground TSV in terms of transmission performance, in *International Conference on Electronic Packaging Technology*, Chengdu, China, 2014, pp. 788–791.

26. A. E. Engin and S. R. Narasimhan, Modeling of crosstalk in through silicon vias, *IEEE Trans. Electromagn. Compat.*, 55(1), 149–158, 2013.

27. T. Song, C. Liu, Y. Peng et al., Full-chip multiple TSV-to-TSV coupling extraction and optimization in 3D ICs, in *Design Automation Conference*, Austin, TX, 2013, pp. 1–7.

28. T. Demeester and D. D. Zutter, Quasi-TM transmission line parameters of coupled lossy lines based on the dirichlet to neumann boundary operator, *IEEE Trans. Microw. Theory Technol.*, 56(7), 1649–1660, 2008.
29. E. X. Liu, E. P. Li, W. B. Ewe et al., Compact wideband equivalent-circuit model for electrical modeling of through-silicon via, *IEEE Trans. Microw. Theory Technol.*, 59(6), 1454–1460, 2011.
30. L. Tsang and X. Chang, Modeling of vias sharing the same antipad in planar waveguide with boundary integral equation and group T-Matrix method, components, *IEEE Trans. Comp. Packag. Manuf. Technol.*, 3(2), 315–327, 2013.
31. Y. J. Zhang, A. R. Chada, and J. Fan, An improved multiple scattering method for via structures with axially isotropic modes in an irregular plate pair, *IEEE Trans. Electromagn. Compat.*, 54(2), 457–465, 2012.
32. X. X. Gu, B. Wu, M. Ritter et al., Efficient full-wave modeling of high density TSVs for 3D integration, in *Proceedings of Electronic Components and Technology Conference*, Las Vegas, NV, 2010, pp. 663–666.
33. Z. H. Guo and G. W. Pan, On simplified fast modal analysis for through silicon vias in layered media based upon full-wave solutions, *IEEE Trans. Adv. Packag.*, 33(2), 517–523, 2010.
34. J. Li, X. C. Wei, and E. P. Li, Analysis of multiple vias coupling in silicon interposer by using cylindrical mode expansion method, in *General Assembly Scientific Symposium*, Beijing, China, 2014.
35. C. R. Paul, *Analysis of Multiconductor Transmission Lines*. John Wiley and Sons: Lexington, KY, 1994.
36. C. D. Wang, Y. J. Cheng, Y. C. Lu et al., ABF-based TSV arrays with improved signal integrity on 3-D IC/interposers: Equivalent models and experiments, *IEEE Trans. Comp. Packag. Manuf. Technol.*, 3(10), 1744–1753, 2013.

Chapter 5

Printed Circuit Board-Level Electromagnetic Compatibility Design

Electromagnetic compatibility (EMC) control and design are critical for the first passing success of all electric products under the strict requirement by international EMC testing standards. The testing frequency becomes much higher than before due to people's never-ending demand for the high-speed communication and high-performance computing. This makes the EMC control and design a great challenge, especially when the testing frequency is above gigahertz. At high frequencies, most EMC control methods are frequency dependent, and it is hard to get a wide-band EMC solution. At the same time, for any structure in high-speed circuits, there is often conflict between signal integrity (SI), power integrity (PI), and electromagnetic interference (EMI) requirements. Therefore, we should consider the balance between different EMC targets. All of these make the sample rules of thumb difficult to be implemented at high frequencies, and most of the EMC problems should be considered and solved case by case. This chapter will focus on three common PCB-level EMC controls: decoupling capacitor (Decap)/shorting vias placement, common mode filter (CMF), and PCB-embedded structures/materials.

1. In order to reduce the power–ground plane (PGP) noise propagation, the impedance of the PGPs is required to be as small as possible. To achieve that, what practical EMC engineers usually do is to add Decap/shorting vias anywhere as long as there is room left on the PCBs. This is sometimes overdesigned. We will briefly introduce the equivalent circuit of Decap, followed by analysis of the local shielding effect of the Decap/shorting via. After that, based on the modal field distribution of the PGPs proposed in Chapter 2, we will analyze and optimize the layout of the through-hole vias, in order to reduce the unwanted coupling between different devices which are mounted on the same PGPs.

2. Differential lines are widely used for high-speed Serializer–Deserializer circuits. They have high immunity to EMI and low radiation emissions. Unfortunately, there is always CM noise on the differential lines, which is due to the asymmetric structures and discontinuities in practical PCBs. From the view of the transmission line matrices, we present the relationship between the CM, the DM, and the single-ended mode. We also discuss the CMF designs to suppress the CM noise. A meander line–quarter wavelength resonator hybrid CMF is proposed [1]. A simple and accurate transmission line model is employed to describe the stopband behavior of the proposed structure. The resonant frequencies and harmonics of quarter wavelength resonators are employed to form stopbands of the CM. These stopbands together with the stopbands of the meander line are connected in series, which are useful to expand the CM suppression bandwidth. Meanwhile, compensation strips are proposed to reduce the insertion loss of the DM. A reduction of 15 dB of CM propagation above 0.8 GHz is obtained by using this hybrid structure. The measurement results of PCBs, which are designed and fabricated according to the simulation model, show good correlation with the simulation results.

3. Due to their compact size and small parasitic *RLCG* than conventional lumped and surface-mounted components, PCB-embedded structures/materials attract more attentions for EMC control, such as the embedded decoupling capacitors. In this chapter, we talk about two kinds of PCB-embedded structures/materials: the embedded filters [2] and absorbers [3]. At the passband of the embedded filter, it provides a path for the return current of the signal. Therefore, it prevents the propagation of the digital switching noise along the power distribution network (PDN) and improves the PI. The isolation band is tunable by adjusting the dimensions of the filter. Next, the embedded absorbing materials are placed on different locations between power and ground planes to compare their performance of noise reduction.

5.1 Reduction of PGP Impedances

5.1.1 Decoupling Capacitors

Decoupling capacitors are widely used in the PDN of high-speed circuits. The voltage level provided by the PND should keep constant for the normal work of circuits. However, when the circuits draw a heavy current, the power supply cannot respond to that current increasing instantaneously. This results in the supply voltage of the circuit decreases until the power supply responds. The Decap works as the circuit's local spare battery. It is placed between the power and ground planes/grids, and as close as possible to the circuit requiring the heavy current. Decap is fully charged when there is no current change from the circuits. When a heavy current is drawn from the circuit, the Decap can provide the spare current immediately through its discharging. This is helpful to reduce the voltage fluctuation along the PDN.

From the view of the frequency domain, the ideal Decap can be taken as a short circuit at high frequencies. When it is placed between the power and ground planes, it can provide a conductive path for the high-frequency return current. It can be used to control the flowing of the return current, and then reduce the input impedance of the PGPs. At the same time, because it can reduce the electric field propagating between the power and ground planes, it also works as a "shielding barrier" to eliminate the electromagnetic noise propagation inside the PDN.

The lumped Decap has the parasitic inductance and resistance. The parasitic inductance prevents the Decap providing the spare current immediately, and also increases the Decap's impedance at high frequency. Due to the parasitic inductance, the application of the lumped Decap is limited to frequencies less than hundreds of megahertz. The Decap can be taken as the series of its capacitance, equivalent series inductance (ESL) and equivalent series resistance (ESR). The equivalent circuit of the Decap is shown in Figure 5.1. ESL is related to the pins of the Decap, and ESR is related to the leakage current of the Decap. Table 5.1 lists the typical values of capacitance, ESL, and ESR for one kind of commercial Decaps.

Figure 5.1   Equivalent circuit of the decoupling capacitor.

Table 5.1 Typical Values of C, ESL, and ESR for One Kind of Commercial Decaps

C (nF)	100	10	1	0.1
ESL (nH)	0.42	0.45	0.41	0.55
ESR (Ω)	0.016	0.078	0.279	0.227
Resonant frequency (MHz)	14.42	45.04	177.8	433.4

The total Decap impedance can be written as

$$Z_C = \text{ESR} + j\omega\text{ESL} + \frac{1}{j\omega C}$$

Decap has the resonant frequency $f_{res} = 1/2\pi\sqrt{\text{ESL}\cdot C}$. At this resonant frequency, the Decap magnitude gets its minimum value as ESR. For frequencies below f_{res}, the Decap is capacitive, whereas for frequencies above f_{res}, the Decap is inductive. Figure 5.2 shows the change of the magnitude of the Decap impedance with the frequency. For the Decap used to reroute the high-frequency return current in the PGPs, we hope the magnitude of its impedance can be smaller enough. In this way, the highest working frequency of Decap is decided by the ESL.

Figure 5.3 shows the PGPs with one port. The substrate has a thickness of 1.2 mm, a relative permittivity of 4.4, and a loss tangent of 0.02. The Decap of Figure 5.2 is added between the power and ground planes, and is very close to the port. Now the input impedance at the port is the parallel connection of Decap

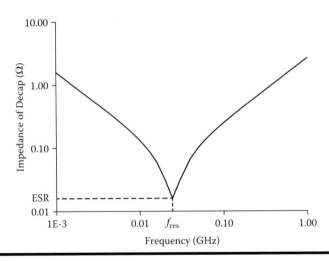

Figure 5.2 Impedance of the decoupling capacitor, where C = 100 nF, ESL = 0.42 nH, and ESR = 0.016 Ω.

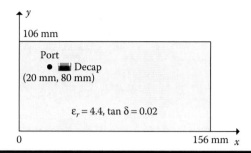

Figure 5.3   PGP with one port and Decap, where the substrate thickness is 1.2 mm.

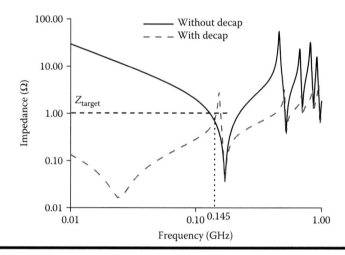

Figure 5.4   Magnitude of the input impedance with and without the Decap.

impedance and Z_{ii}, where Z_{ii} is the input impedance of the PGPs without the Decap, of which the value can be calculated from Chapter 1. Figure 5.4 shows the magnitude of the input impedance with and without the Decap. After adding the Decap, the total capacitance of the PGPs is increased. This result in the magnitude of the input impedance is reduced and the first series resonant frequency shifts to a lower frequency. If the target impedance Z_{target} is defined in Figure 5.4, the largest working frequency of the Decap can be read from the figure as 145 MHz.

5.1.2   Local Shielding of Decaps/Shorting Vias

For the PI of the PGPs in multilayered PCBs, the Decap is placed near to the power–ground lines as shown in Figure 5.5a, so it can reduce the input impedance Z_{in} at the ends of the power–ground lines. The Decap or shorting via also can be used for the SI of the high-speed signal passing through the PGPs as shown in

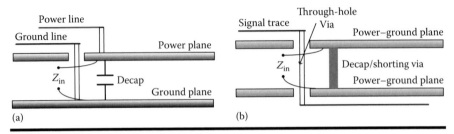

Figure 5.5 Local shielding effects of Decap/shorting via for (a) PI and (b) SI, where the cross sections are shown.

Figure 5.5b, where the Decap/shorting via is placed near the through-hole via and provides a high-frequency conducting path for the return current of the signal. It also reduces the input impedance Z_{in} of PGPs shown in Figure 5.5b. We refer this as the *local shielding*, because in such application the Decap/shorting via is very close to the object, where a low input impedance is required. It should be noted that the Decap/shorting via shown in Figure 5.5 can not only reduce the input impedance but also reduce the mutual coupling between the power lines/through-hole via with other components on the same PGPs. It can shield the electromagnetic wave propagation to and from the power lines/through-hole via.

In the following, the placement of Decap/shorting via for local shielding is studied. For simplify, only the shorting via is considered, and the obtained conclusion can be easily extended to the Decap. The input impedance is simulated for the cases without the shorting vias and with the shorting vias placed at different locations around the port. The port is defined where a low input impedance is required. First, let us consider the case where the port is far away from the edges of the PGPs as shown in Figure 5.6a. One shorting via is placed near the port, and Via1–Via4 present the four different locations of this shorting via. The distances between the port and Via1–Via4 are the same, which is 2 mm. The simulated input impedance at the port is plotted in Figure 5.6b, where the simulation results are obtained by using

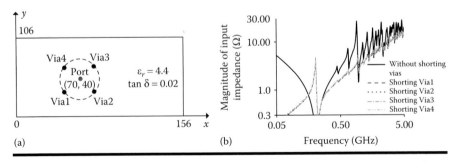

Figure 5.6 (a) The shorting via around the port when the port is far away from the edges of the PGPs (unit: mm), and (b) the input impedance at the port.

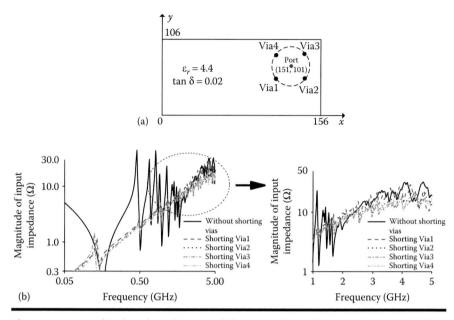

Figure 5.7 (a) The shorting via around the port when the port is near the edges of the PGPs (unit: mm), and (b) the input impedance at the port.

the methods shown in Chapter 2. From Figure 5.6b, we can see that after applying the shorting via, the input impedance at the port is reduced. The shielding effects of Via1–Via4 are almost the same. This means when the port is far away from the edges of the PGPs, the local shielding effect of the Decap/shorting via is isotropic.

Next, let us consider the same PGPs of Figure 5.6a but with the port close to the edges of PGPs, as shown in Figure 5.7a. The distances between the port and Via1–Via4 are still 2 mm. Figure 5.7b shows the simulated input impedance at the port. From this figure, we can see that the shielding effects of Via1–Via4 are anisotropic. Vias3 that is close to the corner of the PGPs can give lower input impedance than Vias at other locations. This suggests that when the port is close to the boundary of the PGPs, the best position of the Decap/shorting via is the position close to the corner or edges of the PGPs.

5.1.3 Global Layout of Signal Traces

According to Chapter 2, for rectangular PGPs with the dimension of $a*b*d$, the mutual impedance z_{ij} between port i and port j can be written as

$$z_{ji} = \frac{j\omega\mu d}{ab} \sum_{m,n=0}^{\infty} \frac{(2-\delta_m)(2-\delta_n)\cos\left(k_x x_i\right)\cos\left(k_y y_i\right)\cos\left(k_x x_j\right)\cos\left(k_y y_j\right)}{k_x^2 + k_y^2 - k^2} \tag{5.1}$$

where δ_m and δ_n are the Kronecker delta with

$$\delta_m = \begin{cases} 1, & \text{if } m = 0 \\ 0, & \text{if } m \neq 0 \end{cases} \tag{5.2}$$

$$\delta_n = \begin{cases} 1, & \text{if } n = 0 \\ 0, & \text{if } n \neq 0 \end{cases} \tag{5.3}$$

$k_x = m\pi/a, k_y = n\pi/b, m, n = 0, 1, 2, \ldots$, and (x_i, y_i) and (x_j, y_j) are the locations of port i and port j, respectively.

The PGPs look like a parallel-plate waveguide, through which the electromagnetic noise propagates and is coupled between different vias passing through the same PGPs. This results in SI issues. z_{ij} in Equation 5.1 is used to calculate this unwanted coupling. It will approach infinite at the resonant frequencies f_{mn} of the PGPs:

$$f_{mn} = \frac{150}{\sqrt{\varepsilon_r \mu_r}} \sqrt{\left(\frac{m}{a}\right)^2 + \left(\frac{n}{b}\right)^2} \text{ GHz, } m, n = 0, 1, 2, \ldots.$$

From Equation 5.1, we can see when port i or port j located at some *zero regions*, z_{ij} will be zero, and there will be no noise coupling between port i and port j. Those zero regions can be obtained from the modal electric field distribution of the PGPs. For the rectangular PGPs, the (m, n) mode is $\cos[(m\pi/a)x]\cos[(n\pi/b)y]$, which is plotted in Figure 2.7. In Figure 2.7, the green regions present the zero regions. Let $\cos[(m\pi/a)x]\cos[(n\pi/b)y] = 0$. We can find the zero regions for (m, n) mode as

$$x = \frac{k + 0.5}{m} a, 0 \leq k \leq m - 1, \text{ for } m \neq 0 \tag{5.4}$$

$$y = \frac{l + 0.5}{n} b, 0 \leq l \leq n - 1, \text{ for } n \neq 0 \tag{5.5}$$

Figure 5.8 shows the zero regions of a rectangular PGP for $(1, n)$, $(m, 1)$, $(2, n)$, and $(m, 2)$ modes. When we want to reduce the coupling between two signal traces passing through the same PGP, we should do the layout optimization of their through-hole vias. For example, choose port 1 and port 2 locations in Figure 5.8. Port 1 is "invisible" for modes $(1, n)$ and $(m, 1)$, and port 2 is "invisible" for modes $(2, n)$ and $(m, 2)$. In this case, the strong coupling due to the PGP's resonance can be greatly reduced.

The mutual impedances of the PGP before and after the layout optimization are shown in Figure 5.10. The PGP shown in Figure 5.9 is used for this comparison.

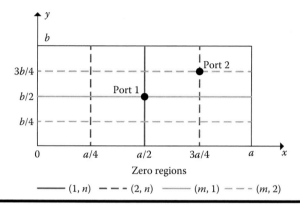

Figure 5.8 **Zero regions of a rectangular PGP for (1, *n*), (*m*, 1), (2, *n*), and (*m*, 2) modes.**

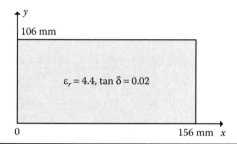

Figure 5.9 **PGPs with the substrate thickness of 1.2 mm.**

The locations of port *i* and port *j* are (59 mm, 39 mm) and (137 mm, 90 mm), respectively, before the layout optimization, and (78 mm, 53 mm) and (117 mm, 79.5 mm), respectively, after the layout optimization. From Figure 5.10, we can see that by the layout optimization, we can reduce z_{ij} in the frequency band about 250 MHz–1.5 GHz. This is helpful for the layout design of signal traces and vias when we want to isolate the noisy devices and the sensitive devices that are connected to the same PGP.

Equations 5.4 and 5.5 of the layout optimization are based on the rectangular PGP. For PGPs with arbitrary shapes, there is no analytical formula for the zero regions. However, commercial software such as SIWave (from ANSYS company) can be used to calculate the modal electric field distributions, and then the zero regions of each (*m*, *n*) mode can be obtained. Following the same layout optimization idea in this section, we also can get the best positions of ports that are *invisible* to the first several resonances of the PGP. The first several resonant modes of the PGP are often important for most of the current electronic devices, because those resonant frequencies of high-order modes are usually beyond the working frequency.

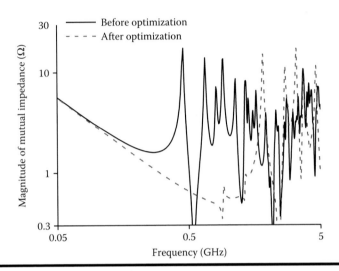

Figure 5.10 Mutual impedances of the PGPs before and after the layout optimization.

5.2 CM Filter

Nowadays, in high-speed digital circuits, signals are generally transmitted by differential signaling schemes [4] because of their benefits of improved SI, low EMI, and high immunity to noise. However, in practical circuits, the common mode (CM) noise due to the timing skew or amplitude unbalance along the differential signal paths is unavoidable and the CM noise above gigahertz frequency will exacerbate EMI problems and degrade SI as well as PI in high-speed digital circuits. Therefore, the maximum CM voltage limitation is considered in specification of high-speed gigahertz differential mode (DM) transmission systems, such as USB 3.0. In addition, as high-speed systems generally operate in several gigahertz bandwidths currently, CMF with enough CM suppression and a satisfactory SI performance in ultrawide frequency band is drawing a wide attention.

Researchers have proposed some methods to reduce or eliminate the CM noise on differential lines. A CM choke is widely used in EMC engineering [5,6]. However, this approach is less effective at frequencies above gigahertz due to its parasitic elements. A defected ground structure provides an interesting method to design a CMF on a printed circuit board (PCB). A defect ground structure (DGS) can suppress the CM noise while having less effect on the differential signal transmission, and it does not consume too much PCB area [7–10]. Unfortunately, a DGS is a narrowband filter because it is based on the single resonant frequency. Other resonant structures are also proposed to design a CMF [11]. However, they still suffer from a narrow resonant band. As a wideband filter, an electromagnetic bandgap (EBG) is proposed to eliminate the CM [12–14], but needs to etch the whole ground plane and has the potential to degrade the SI. In [15] and [16], a new structure is proposed and optimized to suppress the differential-to-common conversion by using tightly coupled microstrip lines. That structure is specially designed to compensate the bend discontinuities and shows a wide bandwidth.

5.2.1 CM and DM

In this section, the voltage, current, and characteristic impedance of CM and DM propagating along a differential pair are discussed. The multiconductor transmission line theory is used to define these parameters. In the high-speed PCB, to keep a good symmetry, the differential pair is made up of two identical coupled microstrip lines or striplines, where the metal planes above or below them are used as their reference/ground planes. This structure can be equivalent to the two identical and coupled transmission lines with a ground plane as shown in Figure 5.11. For simplicity, we assume the metal and substrate are lossless. $V_{1/2}$ and $I_{1/2}$ are the single-ended voltage and current along these two transmission lines, respectively. l_{11} and l_{12} are the per-unit-length (p.u.l.) parasitic self-inductance and mutual inductance, respectively, and c_{11} and c_{12} are the p.u.l. parasitic self-capacitance and mutual capacitance, respectively.

For the two coupled transmission lines, their transmission line equations are

$$\frac{\partial V(z)}{\partial z} = -j\omega[L]I(z) \tag{5.6}$$

$$\frac{\partial I(z)}{\partial z} = -j\omega[C]V(z) \tag{5.7}$$

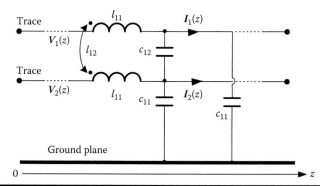

Figure 5.11 Two identical lossless coupled transmission lines with the ground plane.

where:

$$V(z) = \begin{bmatrix} V_1(z) \\ V_2(z) \end{bmatrix}$$

$$I(z) = \begin{bmatrix} I_1(z) \\ I_2(z) \end{bmatrix}$$

The inductance and capacitance matrices are

$$[L] = \begin{bmatrix} l_{11} & l_{12} \\ l_{12} & l_{11} \end{bmatrix} \tag{5.8}$$

$$[C] = \begin{bmatrix} c_\Sigma & -c_{12} \\ -c_{12} & c_\Sigma \end{bmatrix} \tag{5.9}$$

$c_\Sigma = c_{11} + c_{12}$ means the p.u.l. capacitance between one trace and the ground plane when another trace is connected with the ground plane. ($[C]$ is also called Maxwell capacitance matrix in some software, such as 2D Extractor [from ANSYS company].)

For the CM, the two traces are excited at one end by the voltages V_c with equal amplitudes and same directions as shown in Figure 5.12, and they will produce the currents $I_c/2$ with equal amplitudes and same directions flowing along two traces, where V_c and I_c are defined as the CM voltage and current, respectively. For the DM or differential signaling, the two traces are excited at one end by the voltages with equal amplitudes and opposite directions as shown in Figure 5.12, and they

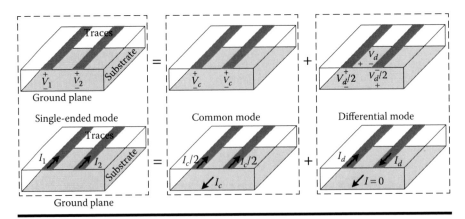

Figure 5.12 Relationship of voltages and currents between the single-ended mode, CM, and DM.

will produce the currents with equal amplitudes and opposite directions flowing along two traces. The voltage between two traces is V_d, which is defined as the DM voltage. The loop current I_d is defined as the DM current, and the total equivalent return current flowing on the ground plane is zero. From Figure 5.12, we can get the transform matrices of voltages and currents between the single-ended mode, CM, and DM as

$$\begin{bmatrix} V_1 \\ V_2 \end{bmatrix} = [\boldsymbol{T}_V] \begin{bmatrix} V_c \\ V_d \end{bmatrix}, \begin{bmatrix} I_1 \\ I_2 \end{bmatrix} = [\boldsymbol{T}_I] \begin{bmatrix} I_c \\ I_d \end{bmatrix} \tag{5.10}$$

where:

$$[\boldsymbol{T}_V] = \begin{bmatrix} 1 & 1/2 \\ 1 & -1/2 \end{bmatrix}, [\boldsymbol{T}_I] = \begin{bmatrix} 1/2 & 1 \\ 1/2 & -1 \end{bmatrix} \tag{5.11}$$

Applying Equation 5.10 into 5.6 and 5.7, we can get

$$\frac{\partial}{\partial z} \begin{bmatrix} V_c(z) \\ V_d(z) \end{bmatrix} = -j\omega[\boldsymbol{L}_m] \begin{bmatrix} I_c(z) \\ I_d(z) \end{bmatrix} \tag{5.12}$$

$$\frac{\partial}{\partial z} \begin{bmatrix} I_c(z) \\ I_d(z) \end{bmatrix} = -j\omega[\boldsymbol{C}_m] \begin{bmatrix} V_c(z) \\ V_d(z) \end{bmatrix} \tag{5.13}$$

where:

$$[\boldsymbol{L}_m] = [\boldsymbol{T}_V]^{-1}[\boldsymbol{L}][\boldsymbol{T}_I] = \begin{bmatrix} l_c & 0 \\ 0 & l_d \end{bmatrix} = \begin{bmatrix} \dfrac{l_{11} + l_{12}}{2} & 0 \\ 0 & 2(l_{11} - l_{12}) \end{bmatrix} \tag{5.14}$$

$$[\boldsymbol{C}_m] = [\boldsymbol{T}_I]^{-1}[\boldsymbol{C}][\boldsymbol{T}_V] = \begin{bmatrix} c_c & 0 \\ 0 & c_d \end{bmatrix} = \begin{bmatrix} 2c_{11} & 0 \\ 0 & \dfrac{c_{11}}{2} + c_{12} \end{bmatrix} \tag{5.15}$$

Because the off-diagonal elements of $[\boldsymbol{L}_m]$ and $[\boldsymbol{C}_m]$ are zeros, Equations 5.12 and 5.13 present two independent equivalent transmission lines, which are related to CM and DM, respectively. The transmission line Equations 5.12 and 5.13 of the CM and DM can be solved independently. Although for the transmission line Equations 5.6 and 5.7 of the single-ended mode, (V_1, I_1) and (V_2, I_2) are dependent on each other.

The impedance design of the differential line is very important for the impedance matching of the differential line with other interconnectors. In the following,

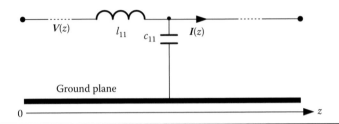

Figure 5.13 Single-ended lossless transmission line.

the propagation constants and characteristic impedances are compared between CM, DM, and single-ended mode. The transmission line of the single-ended mode is defined in Figure 5.13, which has the same p.u.l. self-inductance l_{11} and self-capacitance c_{11} as those in Figure 5.11.

Propagation constants

$$\gamma_c = j\omega\sqrt{c_c l_c} = j\omega\sqrt{c_{11}(l_{11} + l_{12})} \text{ for CM} \tag{5.16}$$

$$\gamma_d = j\omega\sqrt{c_d l_d} = j\omega\sqrt{(c_{11} + 2c_{12})(l_{11} - l_{12})} \text{ for DM} \tag{5.17}$$

$$\gamma_o = j\omega\sqrt{c_{11} l_{11}} \text{ for single-ended mode} \tag{5.18}$$

Characteristics impedances

$$Z_c \equiv \sqrt{\frac{l_c}{c_c}} = \frac{1}{2}\sqrt{\frac{l_{11} + l_{12}}{c_{11}}} \text{ for CM} \tag{5.19}$$

$$Z_d \equiv \sqrt{\frac{l_d}{c_d}} = 2\sqrt{\frac{l_{11} - l_{12}}{c_{11} + 2c_{12}}} \text{ for DM} \tag{5.20}$$

$$Z_o \equiv \sqrt{\frac{l_{11}}{c_{11}}} \text{ for single-ended mode} \tag{5.21}$$

When the distance between two traces is large enough (usually larger than three times of the trace width), the electromagnetic coupling between two traces is so weak that $l_{12} \approx 0$ and $c_{12} \approx 0$. In this case, $\gamma_c = \gamma_d = \gamma_0$, and $Z_c = Z_o/2$, $Z_d = 2Z_o$. When the distance between two traces is decreased, due to the electromagnetic coupling we have $l_{12} > 0$ and $c_{12} > 0$, so $Z_c > Z_o/2$ and $Z_d < 2Z_o$. An empirical formula is provided in [17] to consider the electromagnetic coupling when calculating Z_d.

It should be noted that Equation 5.10 can be taken as the encoding of the single-ended voltage and current. Because the CM and DM can be solved independently, the cross talk between the encoded signal (V_c, I_c) and (V_d, I_d) is greatly reduced compared to the cross talk between the original signal (V_1, I_1) and (V_2, I_2). There is other choice for the encoding. For example, we can choose an orthonormal matrix to be used as $[T_V]$ and $[T_I]$:

$$[T_V]=[T_I]=[T]=\frac{1}{\sqrt{2}}\begin{bmatrix}1 & 1 \\ 1 & -1\end{bmatrix} \tag{5.22}$$

Observe that $[T]^{-1} = [T]^T = [T]$. Applying Equation 5.22 into 5.6 and 5.7, we can get

$$[L_m]=\begin{bmatrix}l_e & 0 \\ 0 & l_o\end{bmatrix}=\begin{bmatrix}l_{11}+l_{12} & 0 \\ 0 & l_{11}-l_{12}\end{bmatrix} \tag{5.23}$$

$$[C_m]=\begin{bmatrix}c_e & 0 \\ 0 & c_o\end{bmatrix}=\begin{bmatrix}c_{11} & 0 \\ 0 & c_{11}+2c_{12}\end{bmatrix} \tag{5.24}$$

For this case, the coupled transmission line can also be decoupled into two independent equivalent transmission lines. In some literatures the modes related to these two equivalent transmission lines are named as even mode and odd mode, respectively. However, sometimes there is confusion between the odd mode and the DM, and the even mode and the CM. From the view of the transmission line, the major difference between them is their different transform matrices $[T_V]$ and $[T_I]$.

Different $[T_V]$ and $[T_I]$ result in decoupled equivalent transmission lines with different p.u.l. parameters. Figure 5.14 shows the p.u.l. capacitances for CM (c_c), DM (c_d), even mode (c_e), and odd mode (c_o). $c_c = 2c_{11}$ is the capacitance between the traces and the ground plane when two traces are connected, $c_d = c_{11}/2 + c_{12}$ is the capacitance between two traces, $c_e = c_{11}$ is the capacitance between one trace and the ground plane when a perfect magnetic wall (open circuit) is inserted between two traces, and $c_o = c_{11} + 2c_{12}$ is the capacitance between one trace and the ground plane when a perfect electric wall (short-circuit) is inserted between two traces. Assuming the two traces are immersed in an effective homogeneous medium characterized by the permeability μ and permittivity ε, the p.u.l. inductance of those modes can be approximated according to $l_i c_i = \mu\varepsilon$ with $i = c, d, e,$ or o.

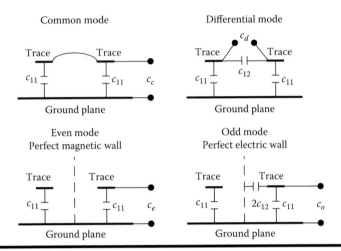

Figure 5.14 p.u.l. capacitances for CM(c_c), DM(c_d), even mode(c_e), and odd mode(c_o) (cross section).

5.2.2 CMF

The CMF is designed according to the different current distributions between the DM and the CM. Figure 5.15a and b shows the current distributions of the DM and CM propagating along the same differential line, respectively. For the DM, the currents flow along two traces with the opposite direction. Their return currents flowing on the ground plane also have the opposite directions. When the distance between the traces and the ground plane is large enough (usually larger than the pitch of the traces), these two return currents will merge together, and then the total current flowing on the ground plane is very small [17]. For the CM, its current follows along the two traces with the same phase, and its return current follows on the ground plane below the traces. The fundamental of the CMF is to eliminate the CM current and allow the DM current. In the following, a defected ground structure-type CMF is employed to demonstrate how the CMF works.

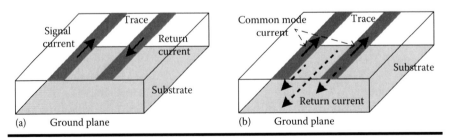

Figure 5.15 Current distributions of (a) the DM and (b) the CM.

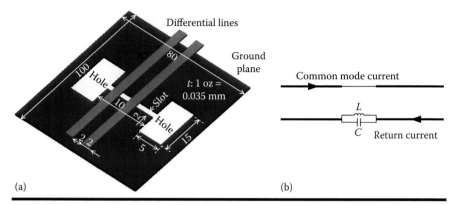

Figure 5.16 **(a) Dumbbell-shaped DGS (unit: mm) and (b) it simplified equivalent circuit.**

The defected ground structure is a cost-effective way to construct the CMF. Figure 5.16a shows a CMF based on the classic dumbbell-shaped DGS. This DGS includes two holes and a slot etched on the ground plane below the differential line. It will change the path of the current following on the ground plane, so that it will greatly influence the CM. Because the induced current on the ground plane of the DM is very small, the DGS does not have much effect on the DM.

For the CM, its return current following on the ground plane has to bypass the DGS. This introduces an additional inductance to the return current. At the same time, the gap of the slot also introduces a capacitance to the return current. Therefore, the dumbbell-shaped DGS can be approximated as an *LC* parallel circuit inserted into the return current of the CM, as shown in Figure 5.16b. At the resonant frequency of this *LC* parallel circuit, the return current of the CM is eliminated and then the CM is suppressed.

Figure 5.17 shows the transmission coefficients of the DM (S_{dd21}) and CM (S_{cc21}) for the dumbbell-shaped DGS shown in Figure 5.16a. From this figure, we can see that the DGS introduces a stopband around 3 GHz to the CM, whereas the DM still can pass the DGS. Figure 5.18a,b plots the current distributions along the differential line for the DM and CM at 3 GHz, respectively. Within the stopband, the *LC* parallel circuit presents a high impedance. Due to the mismatching between the characteristic impedance of the CM (usually 100 Ω) and the *LC* parallel circuit, there is a large reflection at the DGS for the incident CM current. Figure 5.18c plots the return current distribution of the CM on the ground plane at 3 GHz. The reflection of the return current at the DGS can be clearly seen.

Because the CM current has the same phase on the two traces, these two traces can be taken as one single trace when we perform the CMF design. In this way, most of the microstrip/stripline stopband filters proposed in the microwave engineering can be used as the CMF with only a small change. In fact, the dumbbell-shaped DGS also can be employed as a microstrip line stopband filter in RF design. Microwave

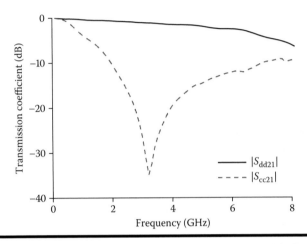

Figure 5.17 $|S_{dd21}|$ and $|S_{cc21}|$ of the CMF based on the dumbbell-shaped DGS.

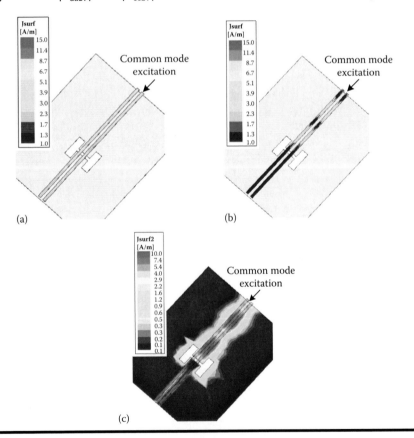

Figure 5.18 Current distributions of (a) the DM and (b) the CM at 3 GHz, and (c) the return current of the CM at 3 GHz.

filters had been studied for a long history, and there had been lots of results published [18], such as Bessel filter, Butterworth filter, and Chebyshev filter. This is helpful for the CMF design. One thing needs to be considered when applying the stopband filters in the microwave engineering to the CMF is that their effects on the DM should be analyzed. In order to reduce their effects on the DM, their structures should be symmetric about the differential lines.

The CMF modeling can be based on different equivalent circuits. The dumbbell-shaped DGS use the LC parallel circuit. The LC series circuit also can be applied between the traces and the ground plane. At the resonant frequency of the LC series circuit, the traces and the ground plane are shorted for the CM. This will also result in large reflection for the CM.

Figure 5.19a shows the magnetic near-field scanning on the surface of a certain microstrip-type differential line, where the differential line is excited by the CM at the left terminal and the testing frequency is 6.59 GHz. The measurement setup is

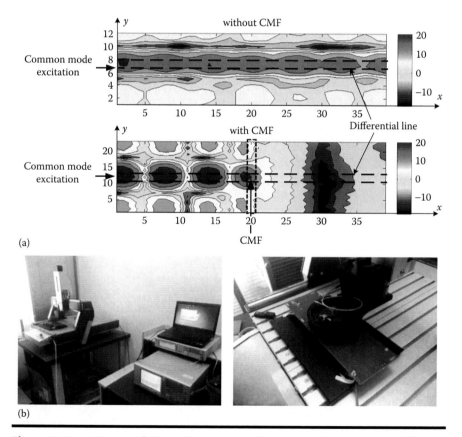

(a)

(b)

Figure 5.19 (a) Scanned H_y on the surface of a certain microstrip-type differential line without and with the CMF, and (b) and the measurement setup.

shown in Figure 5.19b. The differential line is in the center and along the *x*-direction. According to $J_s = \hat{z} \times H$, H_y is scanned and used to present the surface current density along the *x*-direction. The scanned magnetic field shows that the CMF introduces a large reflection of the CM, which is the reason of the CM suppression. This agrees with the conclusion drawn from the above simulations.

5.2.3 Meander Line–Resonator Hybrid Structure

In this section, we explore a meander line–resonator hybrid structure to realize wideband CM suppression, which is more compact in size and inexpensive. We obtain 15 dB of CM noise reduction from 0.8 to 10 GHz, or even higher, which makes it like a low-pass filter. At the same time, in order to keep the DM loss to a small value within the whole wide frequency band, metal compensation strips are proposed with the basic hybrid structure. Finally, PCBs are fabricated and measured to verify the performance of the proposed hybrid structure.

5.2.3.1 Basic Hybrid Structure

In order to obtain a wideband CM suppression filter, a basic meander line–resonator hybrid structure is proposed as shown in Figure 5.20. This structure consists of three metal layers: the differential meander line is on the top layer; the middle layer is the quarter wavelength resonator, of which one end is opened and the other end is connected to the ground plane by a shorting via; a ground plane appears below the quarter wavelength resonator. The width of the quarter wavelength resonator W_r is set to the total width of the differential meander line $2W + S$, as shown in Figure 5.21b, so as to obtain a favorable CM reduction and a compact design.

The differential meander line is widely used on PCBs to reduce CM noise. When it is excited by CM noise, it acts as a coil inductor at low frequencies and

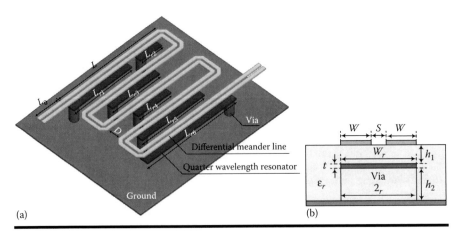

(a)

(b)

Figure 5.20 Basic hybrid structure: (a) three-dimensional view and (b) cross section.

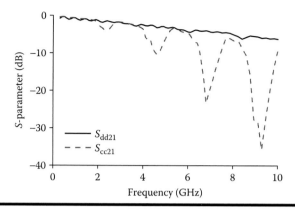

Figure 5.21   Simulation results of the differential meander line.

a stopband filter at higher frequencies, both of which can suppress the CM noise propagation. It has little impact on the differential signal. In addition, this structure consumes small PCB area. Figure 5.21 shows the transmission coefficients of the DM $|S_{dd21}|$ and CM $|S_{cc21}|$ of the differential meander line. The differential meander line has several separated narrow stopbands. In order to broaden its stopband, quarter wavelength resonators are added below it.

Figure 5.22 shows the differential meander line together with one-quarter wavelength resonator below it. The CM current and its return current are also plotted in the figure. The quarter wavelength resonator works like a mask, which prevents the coupling between the differential line and the area donated by A on the ground plane. Hence, the CM return current will flow along the upper surface of the quarter wavelength resonator. This return current is discontinuous at the open end of the resonator, and a displacement current is introduced, which flows through the input impedance Z_{in} of the resonator as shown in Figure 5.22. According to the transmission line theory, Z_{in} approaches infinity at the resonant frequencies of the resonator, which greatly suppresses the return current of the

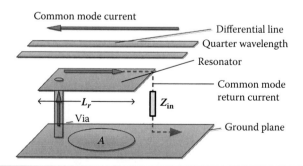

Figure 5.22   CM current's return path on the quarter wavelength resonator.

CM noise, and then the CM propagation. However, Z_{in} would be a small value if the noise's frequency deviates from resonance frequencies, and a CM return current could pass through the resonator easily. In the case of perfect DM signals, because its return current does not go through the ground plane (only when the coupling between each trace and the ground plane is comparable or stronger than that between two traces, the differential signal will induce current on the ground plane), the resonator has little effect on the differential signal.

The resonance frequencies of the resonator are

$$f = (2n+1)c/4L_r\sqrt{\varepsilon_r} \tag{5.25}$$

where:

c is the speed of light in free space
ε_r is the relative dielectric constant of the substrate
L_r is the effective length of the resonator
n presents different resonance frequencies of the resonator (n = 0, 1, 2, 3…)

At these resonance frequencies, the resonator works as a stopband filter, which is then connected to the stopband of the differential meander line in series. Therefore, the whole hybrid structure can expand the stopband of the traditional differential meander line.

Figure 5.23 shows the transmission coefficients of the DM and CM of the differential meander line hybrid with one-quarter wavelength resonator. The length of the resonator is set to 32 mm, which has the primary resonance frequency of 1.1 GHz, and the harmonics of 3.3, 5.7, and 8.0 GHz, as shown in Figure 5.23. Compared to Figure 5.22, besides the original stopbands of the differential meander

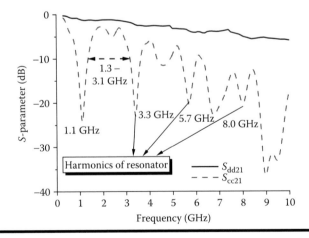

Figure 5.23 Simulation result of differential meander line with one-quarter resonator.

line, this hybrid structure not only forms additional stopbands in the vicinity of 1.1, 3.3, and 5.7 GHz, but also improves the CM reduction near 8.0 GHz. At the same time, compared to Figure 5.22, the transmission coefficient of the DM is not affected by the introduction of the resonator.

From Figure 5.23, we can see that the stopbands are still separated, such as the frequency range from 1.3 to 3.1 GHz, which is still a passband for CM noise. This is because adding one-quarter wavelength resonator to the meander line is inadequate to obtain a wide stopband. Next, more resonators with different lengths are added to the differential meander line to further reduce the value of $|S_{cc21}|$.

By using the resonance frequencies of the resonator, we can choose the initial values of the lengths of the resonators to cover the whole frequency band (0.8–10 GHz), and then full-wave software is employed to fine-tune the lengths of the resonators. The final designed hybrid structure has six resonators, which is shown in Figure 5.20. The dimensions of this basic hybrid structure are as follows: $W = 0.4$ mm, $S = 0.3$ mm, $h_1 = 0.2$ mm, $h_2 = 1.4$ mm, $r = 0.55$ mm, $t = 0.05$ mm, and $D = 2.4$ mm. The PCB material is FR4 with $\varepsilon_r = 4.4$ and a loss tangent of 0.02. The lengths of resonators are $L_{r1} = 20$ mm, $L_{r2} = 8.3$ mm, $L_{r3} = 18$ mm, $L_{r4} = 12$ mm, $L_{r5} = 24$ mm, and $L_{r6} = 32$ mm, respectively. L_0 and L are 24 and 32 mm, respectively. This basic hybrid structure has a CM stopband of –15 dB from 0.8 to 10 GHz or even higher, as shown in Figure 5.24. This makes it like a low-pass filter for CM noise.

CM noise reflects greatly in the CM stopband, as shown in Figure 5.25. This means that a large part of CM transmission converts into CM reflection. This shows the performance of the proposed stopband CMF.

For the proposed hybrid structure, not only the primary resonance frequencies of the resonators but also their harmonics are employed to reduce the CM. By carefully choosing the lengths of the resonators, their primary and harmonic frequencies can cover the whole frequency band.

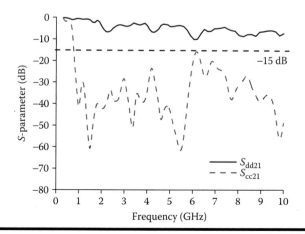

Figure 5.24 Simulation result of the basic hybrid structure shown in Figure 5.20.

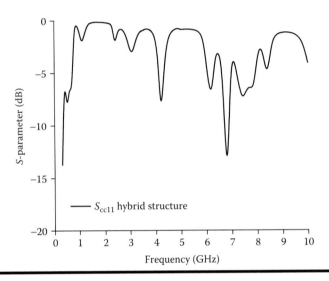

Figure 5.25 $|S_{cc11}|$ **of the hybrid structure without compensation strips.**

Finally, the impact of parameters h_1 (distance between the differential line and the quarter wavelength resonator) and h_2 (distance between the quarter wavelength resonator and the ground plane) on the DM and CM propagations is analyzed. Due to the broadside coupling coefficient is mainly determined by h_1, the CM reduction will greatly change when h_1 is varied from 0.1 to 0.3 mm, as shown in Figure 5.26. The reason for the deterioration of $|S_{dd21}|$, when h_1 is 0.1 mm, is that in this case, the broadside coupling between each trace and

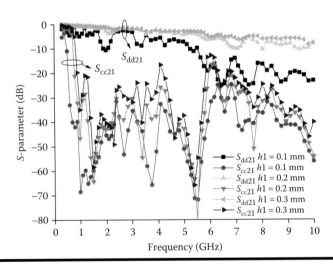

Figure 5.26 **Simulation results of the basic structure with a varying h_1 and a fixed h_2.**

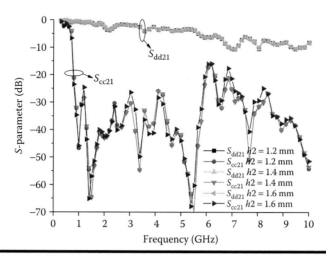

Figure 5.27 **Simulation results of the basic structure with a varying h_2 and a fixed h_1.**

its below-quarter wavelength resonator is comparable or even stronger than the coupling between two traces, which greatly affects the differential current return path. This implies that h_1 should be chosen properly for the balance between CM suppression and DM insertion loss. As for h_2, in this section, it is larger than h_1 and exerts little impact on $|S_{dd21}|$ and $|S_{cc21}|$, as shown in Figure 5.27. Considering the PCB fabrication error and the design requirement of achieving $|S_{cc21}|$ in the range of −15 dB into consideration, h_1 and h_2 are set to 0.2 and 1.4 mm in this section, respectively. For some special applications where h_2 is much small, it will be considered in the further work.

5.2.3.2 Compensation Strips

From Figure 5.24, we can see that although $|S_{cc21}|$ is greatly reduced by adding more resonators, $|S_{dd21}|$ of the basic hybrid structure has some losses in the vicinity of 2, 4, and 6 GHz, in comparison with the $|S_{dd21}|$ of the traditional meander line shown in Figure 5.21 and $|S_{dd21}|$ in Figure 5.23. The reason for these additional losses is that adding several quarter-wavelength resonators in the second layer destroys the integrity and symmetry of meander line structure (every turn of the meander line is accompanied with a resonator with a different length), which may result in discontinuities of the differential lines.

For the parts of differential lines above resonators, there is an additional broadside coupling between each signal trace and the resonators, which will slightly change the characteristic impedance of that part of the differential line, in comparison with other parts of differential lines where there is no resonator under them.

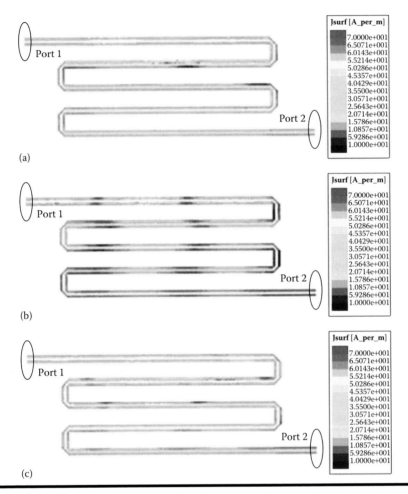

Figure 5.28 Current magnitude distribution excited by DM at the top layer (differential lines): (a) at 5 GHz without compensation strips, (b) at 6.05 GHz without compensation strips, and (c) at 6.05 GHz with compensation strips, where port 1 is excited.

This will result in impedance mismatching losses between different parts of the DM. Figure 5.28a,b shows the current magnitude distribution at 5 and 6.05 GHz (where the DM has an additional loss as shown in Figure 5.24), respectively, where the DM is excited at port 1. We can find that the current shows a larger standing-wave pattern at 6.05 GHz, which results in the additional DM insertion loss at 6.05 GHz in Figure 5.24.

To reduce the DM loss, metal compensation strips are proposed and added below the meander line, as shown in Figure 5.29. These compensation strips are on the same layer of resonators. These compensation strips added to the basic structure

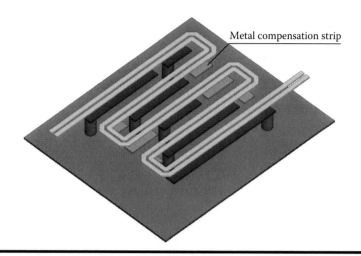

Figure 5.29   Hybrid structure with compensation strips in a three-dimensional view.

make the second layer more integral, which reduce the current standing wave of the DM, as shown in Figure 5.28c. Figure 5.30 shows the comparison of the simulation result between the hybrid structure with compensation strips and that without compensation strips, where except for the compensation strips all dimensions are the same as those used for Figure 5.24. From Figure 5.30, we can see that the

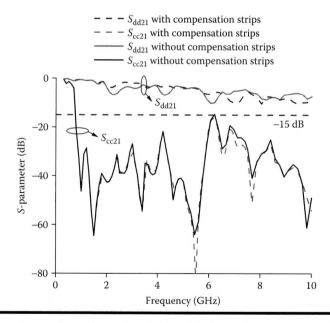

Figure 5.30   Comparison of | S_{dd21} | and | S_{cc21} | between the structure with compensation strips and that without compensation strips.

compensation strips can reduce the DM loss (about 4 dB) in the vicinity of 2, 4, and 6 GHz, and this structure still has a good CM suppression.

5.2.3.3 Measurement Results

In this section, PCBs are fabricated and measured to verify the simulation results of the hybrid structure with compensation strips. Here, SMA connectors are used to connect the fabricated PCB to vector network analyzer (VNA). For the four-port VNA measurement of the differential line, because the pitch of the differential meander line is much smaller than the size of SMA connectors, additional fan-in and fan-out wires must be used to connect the differential line to SMA connectors. This will need a complex de-embedding method to remove the effect of the fan-in and fan-out wires for accurate measurement results. In order to avoid the usage of fan-in and fan-out wires, three measurement models are proposed to measure the transmission coefficients of CM $|S_{cc21}|$, DM $|S_{dd21}|$, and the DM–CM conversion ratio $|S_{cd21}|$, respectively. They are shown in Figure 5.31. The structure in Figure 5.31a, with two shorted ends, is used to measure $|S_{cc21}|$. With two open ends, the structure of Figure 5.31b is used to measure $|S_{dd21}|$. $|S_{cd21}|$ is measured by the third structure in Figure 5.31c, where one end is open and the other is shorted.

Figure 5.32 shows the fabricated PCBs, of which the whole size is 42 mm × 54 mm and other geometrical parameters are the same as those used for Figure 5.30. For the ease fabrication, the blind via of every resonator is replaced by two through-hole vias, which can be seen from Figure 5.32. Figure 5.33 shows the differences between the structure with through-hole vias and that with blind vias. This via replacement is feasible because from Figure 5.33 we can see that $|S_{dd21}|$ is almost the same for through-hole vias and blind vias, and only small changes in $|S_{cc21}|$ are observed.

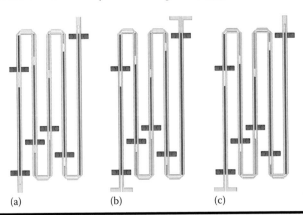

(a) (b) (c)

Figure 5.31 Three measurement models for (a) the transmission coefficients of the CM, (b) the transmission coefficients of the DM, and (c) the DM–CM conversion ratio.

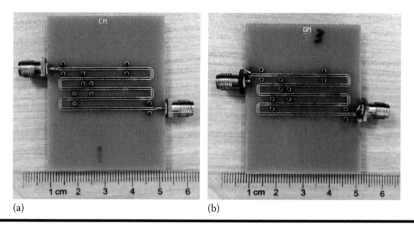

Figure 5.32 Fabricated PCBs for (a) $|S_{cc21}|$ and (b) $|S_{dd21}|$.

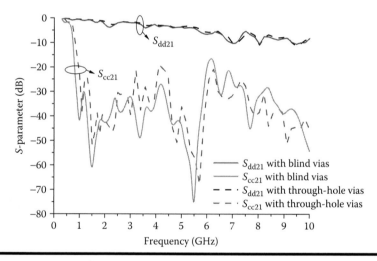

Figure 5.33 Comparison of simulated $|S_{dd21}|$ and $|S_{cc21}|$ between the hybrid structure with through-hole vias and that with blind vias.

The comparisons between measurement and simulation results of $|S_{cc21}|$ and $|S_{dd21}|$ are plotted in Figure 5.34. Good correlation between measurement and simulation can be observed. The error between them is due to the parameter uncertainty of the FR4 substrate and some measurement errors. The measurement results show that a wideband reduction of 15 dB for CM above 0.8 GHz is obtained, and the DM loss is kept to a small value.

Figure 5.35 shows the DM–CM conversion ratio $|S_{cd21}|$ of meander lines and hybrid structure with compensation strips, where we can see that $|S_{cd21}|$ is nearly controlled below –15 dB for the whole band by using the hybrid structure. As we know,

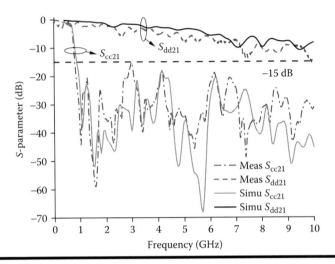

Figure 5.34 Comparison of measured and simulated $|S_{cc21}|$ and $|S_{dd21}|$ of the hybrid structure with compensation strips.

any additional structures added to ideal differential pairs will cause more or less deterioration of symmetry. In this section, the quarter wavelength resonators added to differential pairs will also cause a little increasing of $|S_{cd21}|$. At the same time, $|S_{cd21}|$ of meander lines is quite high as we can see from Figure 5.35. Therefore, optimization on meander lines will be a help to further reduce the $|S_{cd21}|$ of the hybrid structure with compensation strips.

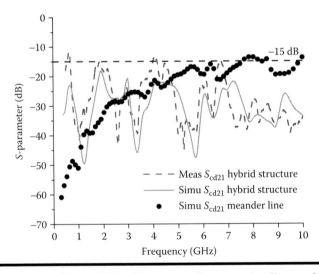

Figure 5.35 Comparison of $|S_{cd21}|$ between the meander line and the hybrid structure with compensation strips.

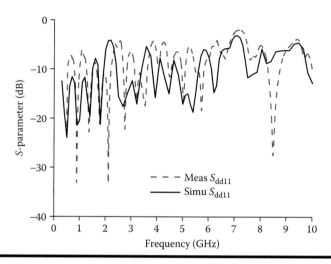

Figure 5.36   Comparison of measured and simulated $|S_{dd11}|$ of hybrid structure with compensation strips.

The simulated and measured DM reflection coefficient results, $|S_{dd11}|$, can be observed in Figure 5.36. This DM reflection is due to the insertion of the quarter wavelength resonators and also the bends of the original meander line.

Figure 5.37 shows the comparison of measured CM transmission coefficients between the traditional meander line structure and the hybrid structure with compensation strips. In comparison with the traditional meander line, further reduction of CM transmission coefficient can be observed by using the proposed hybrid structure.

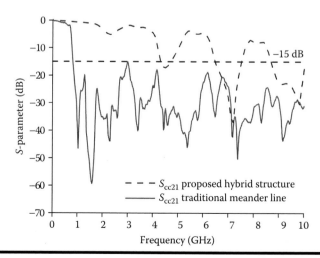

Figure 5.37   Comparison of measured $|S_{cc21}|$ between the traditional meander line and the proposed hybrid structure with compensation strips.

In this section, we propose a compact and wideband meander line–resonator hybrid structure for CM suppression. Approximately 15 dB of CM reduction above 0.8 GHz is achieved. At the same time, the compensation strips are proposed so that the DM insertion loss is kept to a small value. The measurement results of test PCBs give good agreement with simulation results. Both measurement and simulation results show that adding the resonators to the traditional differential meander line can greatly increase its performance for CM suppression.

5.3 PCB-Embedded Structure

5.3.1 PCB-Embedded Filter

The advanced complementary metal–oxide–semiconductor (CMOS) technology requires a stable PDN for the normal work of the high-speed integrated circuits. Usually, the PDN is designed as multilayer power and ground planes to reduce its impedance. As discussed in Chapter 1, when the signals in the horizontal traces flow from one layer to another through a through-hole via, their ground references must change. This results in discontinuities of their return currents, and thus causes signal distortion and noise propagation along the power and ground planes. Things become worse when the noise spectrum includes the inherent resonant frequency of the pair of power and ground planes; in this case, such noise is amplified. When lots of digital circuits simultaneously switch, they draw a heavy current from the power plane to the ground plane. To eliminate the simultaneous switching noise (SSN), a low RF input impedance between the power and ground planes is required.

Several approaches have been proposed to minimize the SSN. The three most frequently used are surface mount decoupling capacitors, isolation, and EBG [19,20]. The surface mount decoupling capacitor fails to provide noise isolation at mid and high frequencies due to its series inductance. The isolation and planar EBG approaches make use of narrow slots etched on the power or ground planes to isolate the noisy circuits from other sensitive circuits and to prevent the propagation of power–ground noise.

In this section, we propose an embedded filter to isolate the SSN. It is different from the available embedded thin-film capacitors [21,22], which require substrates with different materials implemented between the power and ground planes. The thin-film capacitor increases the manufacture cost. The proposed embedded filter does not need new materials integration. One metal layer is introduced between the power and ground planes, which serves as a short path for the return current of the signal at the isolation band. Therefore, it provides a cost-effective method for the noise isolation. At the same time, as an embedding technology, it does not require the power and ground planes to be etched like some planar EBG technologies. Therefore, it does not add SI problem to the signal traces above/below the PGPs.

Based on Chapter one, a simple circuit model is proposed to present the embedded filter, and an efficient de-embedding method is used to extract its admittance. The accuracy of the proposed extraction method is verified by comparing it with the full-wave method.

5.3.1.1 Structure of Embedded Filter

The structure of the proposed embedded filter is plotted in Figure 5.38. Figure 5.38a shows a pair of power and ground planes, where a through-hole via goes through the power plane and is connected to the ground plane. The embedded filter is fabricated between the power and ground planes. It has two multi-finger structures as shown in Figure 5.38b. Two shorting vias are used to connect these two multi-fingers to the power and ground planes, respectively. In Figure 5.38b, l is the length of the fingers and Δ is the gap width between fingers.

For the pair of power and ground planes without the embedded filter, the signal current flowing on the through-hole via will excite a displacement current between the power and ground planes to ensure the continuity of its return current. This displacement current follows from the ground plane, passes the distributed $RLCG$ (resistance, inductance, capacitance, and conductance) between the power and ground planes, and finally arrives at the power plane. It generates the RF noise propagating along the power and ground planes. At the same time, due to the

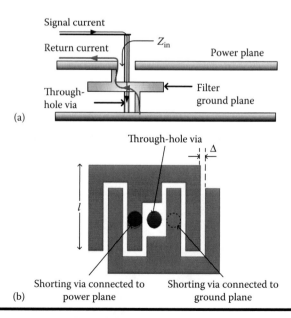

(a)

(b)

Figure 5.38 **(a) Cross section of the pair of power and ground planes with the embedded filter and (b) top view of the embedded filter.**

effect of the distributed $RLCG$ between the power and ground planes, the input impedance Z_{in} in Figure 5.38a is much high at the resonant frequencies of the pair of power and ground planes, which results in the signal distortion.

For the pair of power and ground planes with the embedded filter, at low frequency, the embedded filter is equivalent to a capacitor. However, its capacitance is very small due to its small size. Therefore, it fails to provide a short path for the return current. At high frequency, this embedded filter works as a band-pass filter due to its resonance. At its pass band, its impedance is much low. In this case, it provides a good short path for the return current, as shown in Figure 5.38a. The RF noise produced by the signal current is shielded and the total Z_{in} can be much small. The major difference between this proposed embedded filter and other embedded decoupling capacitors is that we use the structural resonance, instead of the structural capacitance, to reduce Z_{in} at high frequencies.

5.3.1.2 Modeling of Embedded Filter

5.3.1.2.1 Equivalent Circuits

The embedded filter, the through-hole via, and the pair of power and ground planes are represented by their equivalent circuit components as shown in Figure 5.39, where Y_{PG} denotes the input admittance between the power and ground planes with the absence of the through-hole via and embedded filter. Y_{filter} denotes the admittance of the embedded filter. L_{via} and C_{via} denote the parasitic inductor and capacitor of the through-hole via, respectively.

$Y_{in} = 1/Z_{in}$ is the input admittance between the power and ground planes with the presence of the through-hole via and embedded filter. According to Figure 5.39b, it can be written as

$$Y_{in} = j\omega C_{via} + \frac{1}{j\omega L_{via} + \left(1/Y_{PG} + Y_{filter}\right)} \tag{5.26}$$

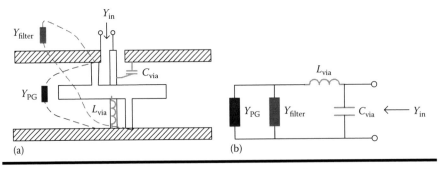

(a)   (b)

Figure 5.39 (a) Equivalent circuits of embedded filter, through-hole via, and the pair of power and ground planes and (b) their connections.

In [23], we proposed a two-dimensional integral equation to extract Y_{PG}. The values of L_{via} and C_{via} can be obtained by the analytical method. Y_{filter} is the only unknown in Equation 5.26. Y_{in} can be obtained after we extract the admittance of the embedded filter.

5.3.1.2.2 Extraction of the Admittance of the Embedded Filter

The effects of the power and ground planes on the coupling between two multi-fingers in Figure 5.38b must be considered during the extraction of Y_{filter}. However, for most available extraction software, such as Q3D (from ANSYS company) and FastCap (from http://www.rle.mit.edu/cpg/research_codes.htm), they are only suitable for the free-space application. Meanwhile, as quasistatic methods, their extracted circuit parameters are frequency independent, which are not accurate at high frequencies. In the following, we propose an efficient de-embedding method to accurately extract Y_{filter}.

As mentioned in [23], there are lots of modes (electromagnetic field distribution) existing between the power and ground planes. They contribute to the electromagnetic coupling between different structures sandwiched between the power and ground planes. Because high-order modes decay very quickly along their propagations, the coupling between the embedded filter and other structures is mainly due to the fundamental mode. This fundamental mode coupling is considered by Y_{PG} in Figure 5.39. (For a general case, Y_{PG} becomes a matrix representing a multiport network.) The above discussion shows that the value of Y_{filter} can be locally calculated. Y_{filter} does not change with the location of the embedded filter.

We can quickly extract the value of Y_{filter} by solving a smaller problem by using the accurate full-wave software HFSS (from ANSYS company). To do so, we reconstruct a similar structure as that in Figure 5.38 with smaller power and ground planes. The absorbing boundary condition is placed along the peripheral of the power and ground planes. The through-hole via and embedded filter are placed at the center of the power and ground planes. From Equation 5.26, we can de-embed Y_{filter} as

$$Y_{filter} = \frac{1}{\left(1/Y_{in}' - j\omega C_{via}\right) - j\omega L_{via}} - Y_{PG}' \qquad (5.27)$$

In the above equation, Y_{PG}' means the input admittance of the reduced power and ground planes with the absence of the through-hole via and embedded filter. Y_{in}' means the input admittance of the reduced power and ground planes with the presence of the through-hole via and embedded filter. Both Y_{PG}' and Y_{in}' can be quickly obtained by using the full-wave HFSS software, which is based on the finite element method. After the value of Y_{filter} is obtained from Equation 5.27, it can be used in the complex power and ground planes with arbitrary shapes.

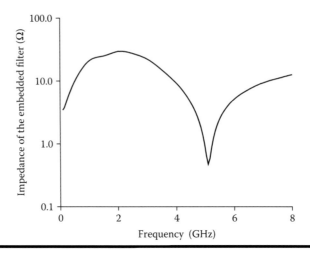

Figure 5.40 The impedance of the embedded filter.

Figure 5.40 shows the extracted impedance of the embedded filter, where $l = 3$ mm and $\Delta = 0.2$ mm. From this figure, we can see that this filter has an isolation band near 5 GHz. Within that isolation band, the impedance is much small, which provides a good short path for the return current.

5.3.1.2.3 Validation of the De-Embedding Extraction Method

The accuracy of the proposed de-embedding extraction method is verified, in which the example under study is the pair of power and ground plane with the embedded filter as shown in Figure 5.41. The substrate has a relative permittivity of 4.4, a loss tangent of 0.02, and a thickness of 1 mm. Figure 5.42 shows the simulated input impedance Z_{in} by using the proposed method and the full-wave HFSS software. Good agreement between the proposed method and the full-wave result can be observed. It should be noted that the proposed de-embedding method is a general extraction method. It can be applied to extract circuit parameters of any embedded structures between the power and ground planes.

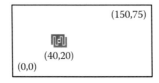

Figure 5.41 A pair of power and ground planes with an embedded filter (unit: mm).

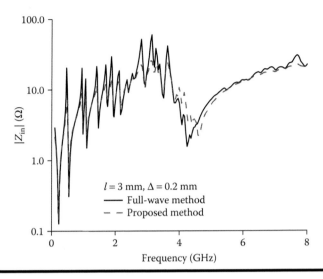

Figure 5.42 Input impedance between the power and ground planes with the embedded filter.

5.3.1.3 Tunable Isolation Band

The isolation band can be easily tuned by adjusting the geometrical dimensions of the filter. Figure 5.43a shows the tuning of the isolation band by changing the finger length l where the gap width Δ is 0.2 mm. Figure 5.43b shows the tuning of the isolation band by changing Δ where l is 3 mm. Here, the power and ground planes under study are shown in Figure 5.41. From these figures, we can see that increasing l or Δ will move the isolation band to lower frequency. In the real application, one can choose different finger lengths and gap widths to achieve low input impedance between the power and ground planes at different frequency bands.

Figure 5.43 also shows that for the power and ground planes without the embedded filter, the trend of the input impedance is to increase with the increasing of the frequency. This is a PI challenge for the ever-increasing working frequency inside the electronic package. The proposed embedded filter stops this trend and pulls down the input impedance at high frequency. At low frequencies (below 2 GHz), the input impedance with the presence of the filter is almost the same as that with the absence of the filter. This is because that at low frequencies, the size of the filter is much smaller than the working wavelength, and it works as a capacitor with a much smaller capacitance. This small capacitance cannot reroute the displacement return current. As we know, the surface mount decoupling capacitors can achieve a larger capacitance, so that it can provide noise decoupling for low frequencies. However, it fails for high frequencies due to its series inductance. In the real applications, the proposed embedded filter can be used together with the surface mount decoupling capacitors to achieve wideband low input impedance.

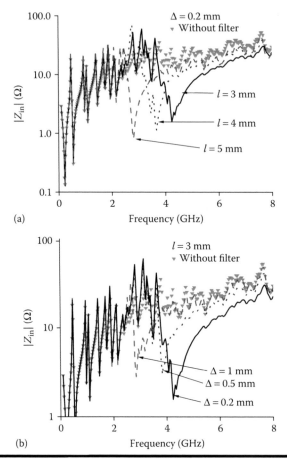

Figure 5.43 **The tuning of the isolation band of the embedded filter by adjusting (a) the length of the finger *l* and (b) the gap width between fingers Δ.**

In this section, we present an embedded filter for the design of the low impedance PDN. The equivalent circuit model of the power and ground pair with the embedded filter is derived for the analysis and design of the embedded filter. An efficient de-embedding method is proposed to extract the admittance of the filter. Through simulation examples, the accuracy of the equivalent circuit model and the tuning of the isolation band are verified. Our future work will be the parameter optimization to achieve much lower input impedance.

5.3.2 PCB-Embedded Absorber

Many kinds of solutions were provided to reduce the noise in the PDN. Among them, most of the researches efforts are spent on the design of EBG. Two popular

EBG structures are the mushroom-like structure and the coplanar structure as shown in Chapter 1 [24–27]. The additional metal layer and the vias connecting the pad to the power or ground plane, which form the mushroom-like EBG structure, will increase the total cost. At the same time, this mushroom structure will be ineffective if the work frequency is above the resonance frequency of the power and ground planes due to the inductive performance of the via. As an improvement, the coplanar-type EBG [26,27] is proposed. Compared with the mushroom-like structure, the coplanar-type EBG has a wider stop bandwidth and saves the cost, but the power or ground plane is periodically etched by slots, which introduces the deterioration of the SI.

In the following, we proposed a method by applying the absorbing material to reduce the impedance of the PDN. In this method, the absorbing material is placed around the power and the ground planes or the via. It will absorb the electromagnetic noise trying to propagate in the PDN. By changing the position and the parameters of the absorbing material, the impedance of the PDN will also change. We do the optimization to improve the performance of this method.

5.3.2.1 Absorbing Material for Noise Reduction in PDN

In this method, we employ the absorbing material to absorb the propagation and reflection of the noise in the cavity-like power and ground planes.

5.3.2.1.1 The Noise Propagation inside the Power–Ground Pair

Figure 5.44 shows a typical power–ground plane pair, in which the interconnect is parallel to the PGPs, and the return current of the signal current in the interconnect is distributed along the nearby power or ground plane. According to the current theory, the signal current together with its return current should form a loop, which means that a displacement current will be induced to ensure the continuity of the return current between the power and ground planes near

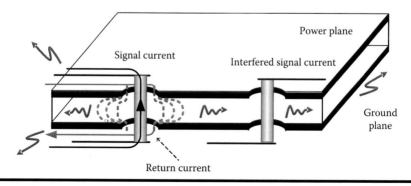

Figure 5.44 Noise coupling inside a power–ground pair.

the via region. This displacement current will excite electromagnetic wave propagation between the power and ground planes. This noise will then be reflected from the boundary of the power and ground planes due to the impedance mismatching between the substrate and the surrounding air. Finally, the noise distribution along the power and ground planes shows a standing-wave pattern. This noise couples to other signal traces passing through the same layer and generates fluctuation of the supply voltage. In the worst case, this noise will be amplified if the noise spectrum covers the inherent resonant frequency of the cavity-like pair of power and ground planes.

5.3.2.1.2 Two Layouts of the Absorbing Material

Our proposed method for reducing the noise between the power and ground planes is based on the theory that the absorbing material could absorb the electromagnetic noise in a wide frequency band if it is properly placed. Figure 5.45 shows the two layouts of our absorbing material, that is, the *boundary-absorber* and the *via absorber*. As shown in Figure 5.45a, the absorbing material is placed along the boundary of the power and ground planes, and Figure 5.45b is the side view. Figure 5.45c,d shows the top view and the side view of the via absorber, where the absorbing material is placed around the via.

In the following, the noise reduction of these two layouts is simulated and compared. In our simulation, the dimension of the power or ground plane is 100 mm × 75 mm, the height of the substrate is 1 mm, and the thickness and the width of the absorbing material are 0.8 and 2 mm separately.

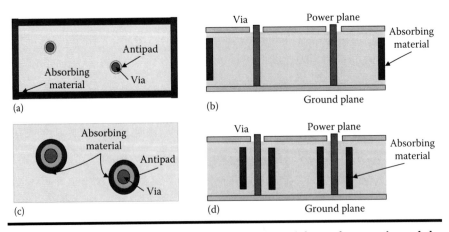

Figure 5.45 The placement of the absorbing material: (a) the top view of the boundary absorber, (b) the side view of the boundary absorber, (c) the top view of the via absorber, and (d) the side view of the via absorber.

5.3.2.2 Validation

In this section, we give the validation of absorbing material for noise reduction in PDN, and the optimization of the parameters and the placement of the absorbing material.

5.3.2.2.1 Electric Loss versus Magnetic Loss

In our study, materials with electric loss and magnetic loss are employed individually to absorb the electromagnetic reflection from the boundary of power and ground planes. The electric loss and magnetic loss show different effects for the noise reduction. We simulate and compare the results of some cases in which the electric loss $\tan \delta_d$ and the magnetic loss $\tan \delta_m$ have different values, and $\varepsilon_r = 10$, $\mu_r = 7$.

Figure 5.46 plots the simulated input impedance in which the materials with different electric loss and magnetic loss are placed along the boundary of the power and ground planes, which is the boundary absorber shown in Figure 5.45a,b. For the reason that the thickness of the substrate is much smaller than the wavelength of the electromagnetic waves, the power and ground planes is approximately regarded as perfect magnetic conductor wall. From the simulation results in Figure 5.46, we conclude that electric loss or magnetic loss alone is not a good choice to get a good performance. This figure shows that after adding the absorbing materials, the cavity

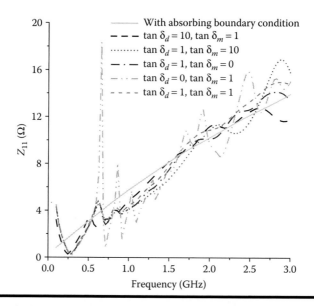

Figure 5.46 The input impedance of the PDN, where $\tan \delta_d$ and $\tan \delta_m$ have different values.

resonance of Z_{11} is greatly removed. However, for all curves in Figure 5.46, Z_{11} always increases with the frequency. This is because that at higher frequencies, the power and ground plane pair becomes an equivalent inductor instead of a capacitor as that at lower frequencies. From the curves with $\tan\delta_d = 1$, $\tan\delta_m = 0$ and $\tan\delta_d = 0$, $\tan\delta_m = 1$, we also can see that the magnetic loss is less effective than the electric loss for the reduction of the input impedance. In the following, we will focus on the different layouts of electrically lossy materials.

5.3.2.2.2 Optimization of the Parameter

From the above simulation results, we know that a larger $\tan\delta_d$ is more useful to reduce the impedance of the PDN than a larger $\tan\delta_m$. The next step is to find out an optimal $\tan\delta_d$.

Figure 5.47 shows the input impedance of the PDN with different values of $\tan\delta_d$ in the boundary absorber layout. In Figure 5.47, we also plot the input impedance in which the boundary of the power and ground planes is placed with the ideal absorbing boundary condition, which is used as a reference to present the best case of such boundary absorbing. Although the bigger the value of $\tan\delta_d$, the smaller the input impedance, from Figure 5.47 we can see that there is an optimal value for $\tan\delta_d$. When $\tan\delta_d$ is above this optimal value, the change of the input impedance becomes small.

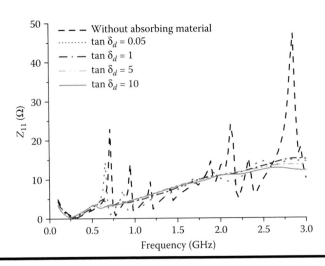

Figure 5.47 **The input impedance of the PDN with different values of $\tan\delta_d$.**

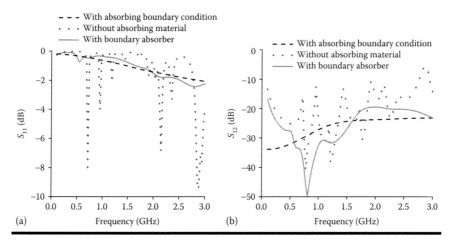

Figure 5.48 **A comparison with three conditions: with absorbing boundary, without absorbing material, and with boundary absorber.**

5.3.2.2.3 Validation of the Boundary Absorber

Through previous simulations, we choose $\varepsilon_r = 10$, $\mu_r = 7$, $\tan\delta_d = 10$, and $\tan\delta_m = 1.5$ as the optimal parameters for the absorbing material placed in the boundary absorber layout. Figure 5.48 plots the simulation results together with the results with/without the ideal boundary condition. Figure 5.48a shows that the reflection is almost matched together with the one with the ideal absorbing boundary condition. And a smaller insertion loss is shown in Figure 5.48b at the range from 0.6 GHz to 1.6 GHz.

5.3.2.2.4 Boundary Absorber vs Via Absorber

By now, we are applying the boundary absorber all along; next, we will simulate the performance of the via absorber shown in Figure 5.45c,d. From the simulation results shown in Figure 5.49, we found that the performance of the via absorber is not as better as that of the boundary absorber at low frequency, ranging from 0.1 to 3.4 GHz. However, as the frequency increases, the via absorber displays a wonderful performance than the boundary absorber as shown in Figure 5.49. We could find that the insertion loss is reduced by 30 dB during 6 to 10 GHz.

In this section, we propose a method by applying the absorbing material in the PDN to mitigating the noise propagating between the power and ground planes. The simulation results show that the absorbing material is able to reduce the noise in the electrical package or the PCB without degrading the SI of the signal traces. The boundary absorber and the via absorber layouts perform well in different frequency ranges.

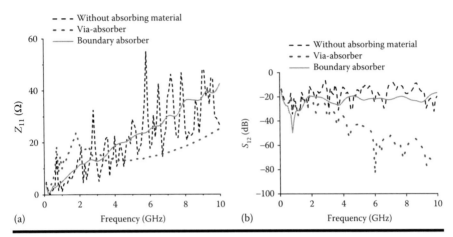

Figure 5.49 **The performance comparison between the boundary absorber and the via absorber: (a) Z_{11} and (b) S_{12}.**

References

1. Y. F. Shu, X. C. Wei, X. Q. Yu, Y. S. Li, and E. P. Li, A compact meander line-resonator hybrid structure for wideband common mode suppression, *IEEE Trans. Electromagn. Compat.*, 57(5), 1255–1261, 2015.
2. X. C. Wei and E. P. Li, Design and modelling of a cost-effective embedded filter for low-impedance power distribution network, *IEEE International Conference on Electrical Performance and Electronic Packaging and Systems*, Austin, TX, October 25–27, 2010.
3. L. S. Zhang, X. C. Wei, M. Ni, and E. P. Li, A novel method for low impedance design of power and ground planes, *IEEE Electrical Design of Advanced Package & Systems Symposium*, Hangzhou, China, 12–14, 2011.
4. J. Fan, X. Ye, J. Kim, B. Archambeault, and A. Orlandi, Signal integrity design for high-speed digital circuits: Progress and directions, *IEEE Trans. Electromagn. Compat.*, 52(2), 392–400, 2010.
5. N. Oka, K. Misu, S. Yoneda, S. Saito, and S. Nitta, A common mode noise filter for high speed and wide band differential mode signal transmission, *10th International Symposium on Electromagnetic Compatibility*, York, UK, 2011.
6. S. Saito, T. Kobayashi, and S. Nitta, High speed signal transmission compatibility with noise suppression by common mode chokes, *10th International Symposium on Electromagnetic Compatibility*, York, UK, 2011.
7. W. T. Liu, C. H. Tsai, T. W. Han, and T. L. Wu, An embedded common-mode suppression filter for GHz differential signals using periodic defected ground plane, *IEEE Microw. Wireless Compon. Lett.*, 18(4), 248–250, 2008.
8. Y. S. Li, X. C. Wei, and E. P. Li, An asymmetric spiral-shaped DGS optimized by genetic algorithm for wideband common-mode suppression, *Asia-Pacific Symposium Electromagnetic Compatibility*, Taipei, Taiwan, 2015.
9. S. J. Wu, C. H. Tsal, T. L. Wu, and T. Itoh, A novel wideband common-mode suppression filter for gigahertz differential signals using coupled patterned ground structure, *IEEE Trans. Microw. Theory Tech.*, 57(4), 848–855, 2009.

10. F. X. Yang, M. Tang, L. S. Wu, and J. F. Mao, A novel wideband common-mode suppression filter for differential signal transmission, *IEEE Electrical Design of Advanced Packaging & Systems Symposium*, Bangalore, India, 2014.
11. G. H. Shiue, C. M. Hsu, C. L. Yeh, and C. F. Hsu A comprehensive investigation of a common-mode filter for gigahertz differential signals using quarter-wavelength resonators, *IEEE Trans. Adv. Packag.*, 4(1), 134–144, 2014.
12. F. D. Paulis, L. Raimondo, S. Connor, B. Archambeault, and A. Orlandi, Design of a common mode filter by using planar electromagnetic bandgap structures, *IEEE Trans. Adv. Packag.*, 33(4), 994–1002, 2010.
13. F. D. Paulis, L. Raimondo, S. Connor, B. Archambeault, and A. Orlandi, Compact configuration for common mode filter design based on planar electromagnetic bandgap structures, *IEEE Trans. Electromagn. Compact.*, 54(3), 646–654, 2012.
14. F. D. Paulis, M. H. Nisanci, A. Orlandi, S. Connor, M. Cracraft, and B. Archambeault, Standalone removable EBG-based common mode filter for high speed differential signaling, *IEEE International Symposium on Electromagnetic Compatibility*, Raleigh, NC, 2014.
15. C. Gazda, D. V. Ginste, H. Rogier, R. B. Wu, and D. D. Zutter, A wideband common-mode suppression filter for bend discontinuities in differential signaling using tightly coupled microstrips, *IEEE Trans. Adv. Packag.*, 33(4), 969–978, 2010.
16. C. Gazda, I. Couckuyt, H. Rogier, D. V. Ginste, and T. Dhaene, Constrained multiobjective optimization of a common-mode suppression filter, *IEEE Trans. Electromagn. Compat.*, 54(3), 704–707, 2012.
17. E. Bogatin, *Signal integrity: Simplified*, Prentice Hall PTR, New Jersey, 2003.
18. D. M. Pozar, *Microwave Engineering*, 4th ed, Wiley, New Jersey, 2012.
19. S. Huh, M. Swaminathan, and F. Muradali, Design, modeling, and characterization of embedded electromagnetic band gap (EBG) structure, *Electrical Performance of Electronic Packaging*, San Jose, CA, 2008, pp. 83–86.
20. T. K. Wang, C. Y. Hsieh, H. H. Chuang, and T. L. Wu, Design and modeling of a stopband-enhanced EBG structure using ground surface perturbation lattice for power/ground noise suppression, *IEEE Trans. Microwave Theory Tech.*, 57(8), 2047–2054, 2009.
21. P. Muthana, A. E. Engin, M. Swaminathan, R. Tummala, V. Sundaram, B. Wiedenman, D. Amey, K. H. Dietz, and S. Banerji, Design, modeling, and characterization of embedded capacitor networks for core decoupling in the package, *IEEE Trans. Adv. Packag.*, 30(4), 809–822, 2007.
22. H. Kim, Y. Jeong, J. Park, S. Lee, J. Hong, Y. Hong, and J. Kim, Significant reduction of power/ground inductive impedance and simultaneous switching noise by using embedded film capacitor, *Electrical Performance of Electronic Packaging*, Princeton, NJ, 2003, pp. 129–132.
23. X. C. Wei, E. P. Li, E. X. Liu, and R. Vahldieck, Efficient simulation of power distribution network by using integral equation and modal decoupling technology, *IEEE Trans. Microwave Theory Tech.*, 56(10), 2277–2285, 2008.
24. R. Abhari and G. V. Eleftheriades, Suppression of the parallel-plate noisein high-speed circuits using ametallic electromagnetic bandgap structure, *in IEEE MTT-S International Microwave Symposium Digest*, 2002, pp. 493–496.
25. T. Kamgaing and O. M. Ramahi, A novel power plane with integrated simultaneous switching noise mitigation capability using high impedancesurface, *IEEE Microw. Wireless Compon. Lett.*, 13(1), 21–23, 2003.

26. T. L. Wu, Y. H. Lin, and S. T. Chen, A novel power planes with low radiation and broadband suppression of ground bounce noise using photonic bandgap structures, *IEEE Microw. Wireless Compon. Lett.*, 14(7), 337–339, 2004.
27. T. L. Wu, Y. H. Lin, T. K. Wang, C. C. Wang, and S. T. Chen, Electromagnetic bandgap power/ground planes for wideband suppression of ground bounce noise and radiated emission in high-speed circuits, *IEEE Trans. Microwave Theory and Tech.*, 53(9), 2935–2942, 2005.

Chapter 6

Interposer Electromagnetic Compatibility Design

With the rapid development of the semiconductor technology, traditional device scaling is meeting its physical limitations. Three-dimensional (3D) and two-and-half-dimensional (2.5D) integration technologies with increased package density and improved system performance have been proposed to solve such problems [1,2]. Among them, 2.5D integration with through silicon via (TSV) and interposer is becoming the most promising More-Than-Moore solutions due to its low fabrication cost.

TSV interposer manufactured by etching vias through silicon substrate and filling the vias with metal can be integrated with heterogeneous chips, such as memory, radio frequency (RF), digital, processor, and analogy, realizing high-density packaging and multifunctional system [3–5]. TSV interposer provides shorter interconnection from die to substrate in the vertical direction, thus reducing resistive-capacitive (RC) and conduction loss [6]. Minimized mismatch of the coefficient of thermal expansion between the chip and the interposer ensures a highly reliability of the package [7].

However, for the interposer to be used for high-speed circuits, the characteristic of the lossy silicon has to be considered. This results in large signal insertion loss and cross talk. One way is to design the novel interconnect structure to change the electromagnetic field distribution and reduce their leakage into the silicon substrate; another way is to use the glass interposer and the through glass via (TGV) instead of the silicon interposer and the TSV, because the glass has less loss than the silicon. In this chapter, we will discuss these two issues.

First, we propose a double-shielded interposer design with two metal layers directly contacting to the top and bottom surfaces of the silicon substrate for the high-speed signal propagating along TSVs [8]. To enhance the shielding effects, shallow highly doped silicon is made on both sides of the silicon interposer forming ohmic contact between the metallization layer and silicon. The metallization layer is etched into a meshed pattern to prevent delamination. Based on the imaging method in Chapter 2, we derive the equivalent circuit model of the proposed double-shielded interposer, and the accuracy and efficiency of the equivalent circuit model is verified by the full-wave method. The proposed design can concentrate the electromagnetic field around TSVs, so that it has a lower insertion loss than the conventional ground-signal (GS) structure without shielding. This is especially useful when the low-resistivity silicon is used for the interposer. Finally, some applied guidelines are proposed to strengthen the performance and simplify the silicon fabrication processes.

Next, two kinds of compact waveguide structures based on TGV technology are studied [9]. The full-wave simulation is applied to analyze the transmission characteristics of these structures. The return loss, insertion loss, and electric field distribution results show that the compact waveguides have a good signal transmission performance. The air-filled TGVs are employed in the proposed waveguides. Simulation results show that these air-filled vias can concentrate electromagnetic field within the waveguides, and thus reduce the electromagnetic interference inside the glass interposer.

6.1 Double-Shielded Interposer

For TSV-based 3D integrated circuit to be used in high-speed applications, the lossy characteristic of the silicon has to be considered. Some previous works have been done to suppress such substrate loss and the corresponding electromagnetic compatibility (EMC) issues. In [10], a TSV equalizer was proposed to intend to increase the low-frequency loss to compensate the high-frequency loss. In [11], a program of shielding ground TSVs was proposed to reduce the transmission loss. At the same time, from the aspect of material, the high-resistivity silicon [12–14] and the thick insulator surrounding TSV were also proposed. However, the high-resistivity silicon substrate will increase the cost, and the thicker insulator faces the difficulty of oxidizing the TSV with a high aspect ratio.

In this section, a double-shielded TSV interposer is presented, where the metal layers together with the highly doped layers are placed on double sides of the silicon substrate (denoted by MS). Under the consideration that a large area of metal plane is easy to peel off during the processing, a meshed metal layer is used. Shallow highly doped silicon not only provides ohmic contact but also enhances the shielding effects of the meshed metal layers. The equivalent circuit model is established by using imaging theory proposed in [15]. The proposed equivalent circuit model is efficient and accurate in estimating the input impedance of the proposed double-shielded TSV interposer. The proposed TSV interposer provides a good shielding of the electromagnetic leakage into the substrate, making the interposer "grounded" and the electromagnetic field concentrated around the TSVs when the signal current is flowing through TSVs. Therefore, the design can help to reduce the total electromagnetic field inside the silicon and then reduce the signal insertion loss caused by the lossy silicon. Finally, some useful guidelines are proposed to improve the design flexibility of the MS shielding structure.

6.1.1 Double-Shielded Interposer and Its Equivalent Circuit

6.1.1.1 Double-Shielded TSV Interposer

Differential lines (GS structure) are commonly used for high-speed signal propagation. The typical GS TSV structure is shown in Figure 6.1. For such GS structure, the electromagnetic emission from the redistribution layer (RDL) will go deeply into the lossy silicon and results in a larger signal insertion loss.

To eliminate such media loss and still use the low-cost and low-resistivity silicon, we propose a new TSV interposer structure as shown in Figure 6.2, where the meshed metal layers directly contact with the silicon substrate surfaces without the insulator layer between them. Here the metal layer servers as a shielding to reduce the electromagnetic field penetration into the silicon. In the following, the structure in Figure 6.2 is referred to as metal–silicon (MS) structure.

A power–ground plane is usually designed to have a meshed pattern for easy processing [16]. In the MS interposer design, in order to avoid the delamination of the metal layer from the silicon substrate, the metal layers are etched into meshed patterns. The top view and dimensions of the meshed plane are depicted in Figure 6.3. In the figure, L_w is the line width of the copper grids, and L_p is the distance between two copper grids.

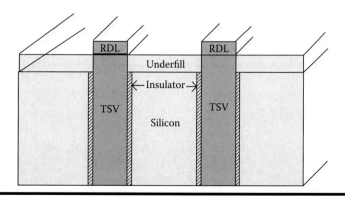

Figure 6.1   Side view of the typical GS TSV pair with RDL.

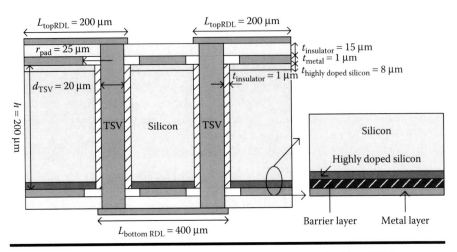

Figure 6.2 **The geometry and design parameters of the MS shielded TSV interposer.**

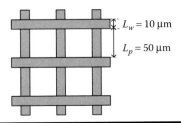

Figure 6.3 **Top view and dimensions of the meshed metal layer.**

For the purpose of making the silicon substrate better shielded, a highly doped silicon epitaxial layer is grown on the silicon surface under the metallization grids, whose doping concentration is bigger than 10^{19}cm^{-3}, as shown in Figure 6.2. The doped silicon forms an ohmic contact between the silicon and metallization grids. The resistivity of such ohmic contact is less than 4.6 μΩm, which provides a better electric contact between the silicon and metallization grids. Such ohmic contact is very popular in real applications, and the highly doped silicon layers are easy to be fabricated [10].

To avoid the copper diffusion into silicon, a zirconium–silicon (Zr–Si) diffusion barrier is adopted between the metallization grids and silicon, acting as the barrier layer and for better adhesion to the silicon. The low ohmic contact of Zr–Si/Si interface with a resistivity of 0.136 μΩm is helpful to keep a low total resistance of metal grids–highly doped silicon. In [17], a thin diffusion barrier of zirconium–nitride (Zr–N) is inserted into the Cu/Si contact system. A zirconium silicide diffusion barrier is made between the Cu/Si interfaces in [18]. Experimental results show that such films have good barrier properties for copper metallization.

6.1.1.2 Equivalent Circuit Model

In this section, the equivalent circuit model of the proposed MS shielded interposer is derived. Figure 6.4 shows the MS shielded interposer without the RDL, where two ports are defined at the top and bottom of the TSV, respectively.

The equivalent circuit model of the MS shielded interposer is proposed in Figure 6.5. The carrier concentration in the highly doped silicon layers is very high, and the contact resistance between barrier Zr–Si and silicon is very low; therefore, the metal grids, barrier layer, and highly doped layer are considered as one ground plane in Figure 6.5. The TSV can be modeled as a series of resistance R_{TSV} and self-inductance L_{TSV}; L_{TSV} is a partial inductance. C_{OX} is the parasitic oxide capacitance between the TSV conductor and the silicon substrate, which can be derived from the model of coaxial cable. The fringing capacitance cross the gap between the TSV and the equivalent ground planes can be represented as C_{gap}. The resistance R_{gap} presents the silicon loss crossing the gap. The values of those parasitic parameters in Figure 6.5 can be obtained as follows:

$$R_{TSV} = \sqrt{(R_{DC})^2 + (R_{AC})^2} \tag{6.1}$$

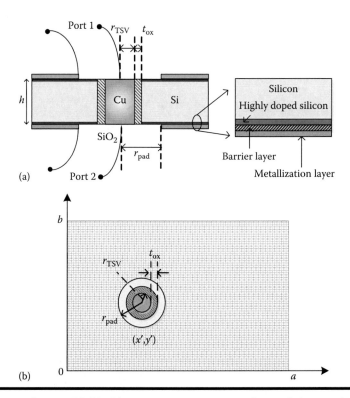

Figure 6.4 The MS shielded interposer: (a) Cross section and (b) top view.

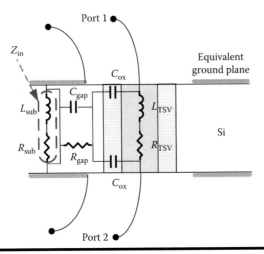

Figure 6.5 **The equivalent lumped circuit model for the proposed MS shielded interposer.**

where R_{DC} and R_{AC} are the DC and AC resistances of the cylindrical TSV, of which the values can be found in [1].

$$L_{TSV} = \frac{55 \, \mu r_{TSV}}{2\pi} \ln\left(1 + 0.01 \frac{h}{r_{TSV}}\right) \tag{6.2}$$

$$C_{OX} = \pi\varepsilon_0\varepsilon_{r,SiO_2} \times \frac{h}{\ln\left(\frac{r_{TSV} + t_{OX}}{r_{TSV}}\right)} \tag{6.3}$$

$$C_{gap} = \frac{4 \, \pi\varepsilon_0\varepsilon_r}{h \ln\left(\frac{r_{TSV} + r_{pad}}{r_{TSV}}\right)} \sum_{n=1,3,5,\ldots}^{2N-1} \frac{1}{k_n^2}\left[1 - \sqrt{\frac{r_{TSV}}{r_{TSV} + r_{pad}}} e^{-k_n r_{pad}}\right] \tag{6.4}$$

where, $k_n = \sqrt{(n\pi/2h)^2 - (2\pi/\lambda_g)^2}$ and $\lambda_g = c/f\sqrt{\varepsilon_r}$. ε_{r,SiO_2} and ε_r are the relative permittivities of the SiO$_2$ and silicon substrate, respectively and μ is the permeability of the silicon substrate

Next, we will derive the formula of Z_{in}, which includes L_{sub} and R_{sub} in Figure 6.5. For the signal current flowing through the TSV, its return current (in the form of displacement current) will flow from one ground plane to the other ground plane in the TSV interposer. The impedance Z_{in} between these two ground planes, that is, the vertical impedance of the MS shielded interposer, should be extracted to accurately calculate this return current. In the following, the imaging method [15] is used to derive the analytical formula of the vertical impedance Z_{in}.

The wave reflection from the boundary of the interposer has contribution to its vertical impedance. In the imaging method, the reflected electromagnetic field by the boundary can be considered as the contributions from the images of the original excitation current, where the boundary of the interposer is taken as the perfect magnetic conductor. By this way, the boundary of the interposer can be removed and an analytical formula of the vertical impedance can be obtained.

In Figure 6.4b, the TSV is located at (x', y'). The vertical impedance of the interposer at (x', y') can be calculated by using the sum of contributions from the original excitation current and all its images as

$$Z_{in} = \frac{\omega\mu h}{4} \sum_{m,n=-\infty}^{\infty} \sum_{\alpha,\beta} H_0^{(2)}\left(k\rho_{mn,\alpha\beta}\right) \tag{6.5}$$

where:

$\rho_{mn,\alpha\beta} = \sqrt{(X_\alpha + 2ma)^2 + (Y_\beta + 2nb)^2}$, $X_\alpha = x' - \alpha x'$, $Y_\beta = y' - \beta y'$, $\alpha/\beta = -1$ or 1
$H_0^{(2)}$ is the second kind Hankel function with the zero order
ω is the angular frequency
k is the complex wave number of the substrate

Because the low-resistivity silicon is employed, the magnitude of $H_0^{(2)}$ $(k\rho_{mn,\alpha\beta})$ decays very quickly when m and n increase. Therefore, when the TSV is not too close to the boundary of the interposer, Equation 6.5 can be approximated as

$$Z_{in} \approx \frac{\omega\mu h}{4} H_0^{(2)}(k r_{TSV}) \tag{6.6}$$

Furthermore, an approximation of the Bessel function is adopted when considering that the TSV radius r_{TSV} is much smaller than the wavelength of interest:

$$H_0^{(2)}(z) \approx \frac{2}{\pi j} \ln z, \quad \text{for small } |z| \tag{6.7}$$

Substituting Equation 6.7 into 6.6, we can rewrite Z_{in} in the following form:

$$Z_{in} = R_{sub} + j\omega L_{sub} \tag{6.8}$$

where R_{sub} and L_{sub} are the resistance and inductance of the MS shielded interposer as shown in Figure 6.5. We can get the frequency-dependent R_{sub} and L_{sub} as

$$R_{sub} = \omega \frac{\mu h}{4\pi} Q, \quad \text{with} \quad Q = -\arctan \frac{\sigma_{Si}}{\omega\varepsilon_0\varepsilon_r} + p\pi \tag{6.9}$$

$$L_{sub} = -\frac{\mu b}{4\pi}[2\ln(\omega r_{TSV}\sqrt{\mu\varepsilon_0\varepsilon_r}) + \ln\sqrt{1 + \tan^2\delta}] \qquad (6.10)$$

where:

p is the minimum positive integer (1, 2, 3...) which meets the condition of $Q \geq 0$

σ_{Si} is the conductivity of the silicon substrate

$\tan\delta = \sigma_{Si}/\omega\varepsilon_0\varepsilon_r$

It should be noted that the formula of Equations 6.9 and 6.10 can be used for the interposer boundary with arbitrary shapes.

In order to validate the efficiency and accuracy of the equivalent circuit model of the MS double shielded interposer, the 3D full-wave simulator is applied to calculate the S parameters of the structure shown in Figure 6.4 up to 50 GHz and compared with the results obtained from the proposed equivalent circuit model. In the 3D full-wave simulation, both metal grids and highly doped layers are modeled. The size of the interposer is 3 mm × 6 mm and the height is 200 μm. The radius of the TSV is 15 μm, surrounded by an insulator with 1 μm thickness. The simulation results are shown in Figure 6.6.

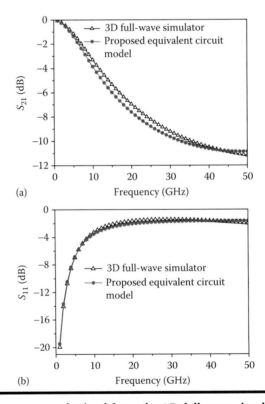

(a)

(b)

Figure 6.6 *S* **parameters obtained from the 3D full-wave simulator and equivalent circuit model: (a)** S_{21} **parameter and (b)** S_{11} **parameter.**

6.1.2 Signal Propagation Analysis

The signal insertion loss of the typical GS structure as shown in Figure 6.1 and the MS double shielded interposer as shown in Figure 6.2 are analyzed and compared in this section. It should be noted that both the two interposers have the same materials and dimensions, except that the MS shielding interposer has the metal layer and shallow highly doped silicon acting as the shielding. The conductivity of silicon (σ_{si}) is 1750 S/m, which is frequency independent during the simulation. The pitch between the signal TSV and the ground TSV in the two interposer structures is 100 μm. The two interposers have the same boundary and excitation conditions. Two lumped ports are defined on the ends of traces: port 1 locates on the left, and port 2 locates on the right.

6.1.2.1 Insertion Loss

Figure 6.7 shows the insertion loss ($-20\log|S_{21}|$) comparison between the MS double shielded interposer (denoted by MS shielding) and the typical GS structure (denoted by typical GS), where the total length of the signal trace is the same for these two structures. It is shown in Figure 6.7 that in the low-frequency range, the insertion losses of the two structures are almost the same. In the low-frequency band, the thickness of the metal plane and shallow highly doped silicon is smaller than the skin depth, which weakens the shielding effects; therefore, the insertion loss of the proposed MS shielded interposer has small distinction with the typical GS structure. However, during the high-frequency range from 25 to 50 GHz, the insertion loss of the proposed MS interposer is smaller than that of the typical GS structure. The conclusion can be drawn that the shielding effect of the MS interposer is obvious at high frequencies.

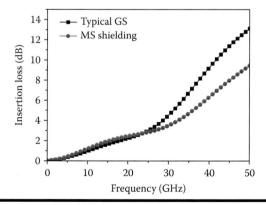

Figure 6.7   Insertion loss comparison between two interposers.

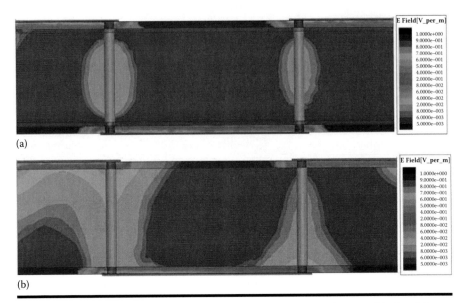

(a)

(b)

Figure 6.8 **Normalized electric field distribution inside the two kinds of inter-poser: (a) MS interposer at 40 GHz and (b) typical GS structure at 40 GHz.**

6.1.2.2 Electric Field Distribution Inside the Interposers

To further analyze the reduction of insertion loss by using the MS interposer, the normalized electric field distribution inside the two structures at 40 GHz is illustrated in Figure 6.8, where the source is applied at the left port, and the radiation boundary is employed.

From the figure, it can be seen that the electric field distribution of the proposed MS interposer is much confined around the TSV. However, for the typical GS structure, the electric field distributions spread in the substrate, which results in the larger signal insertion loss. This implies that the proposed MS interposer with metallization grids directly contacting with the substrate surfaces well "ground" the interposer, which makes the low-resistivity silicon substrate together with the TSV form a "coaxial cable" as shown in Figure 6.8a. At the same time, the highly doped layers in the proposed MS interposer improve the shielding or grounding effect of the metal grids.

It should be noted that the signal loss in the substrate is reduced by changing the structure other than using the high-resistivity silicon, which is helpful to reduce the cost of the packaging.

6.1.2.3 Dielectric and Metal Losses

Finally, we will study and compare the dielectric and metal losses inside the interposer for the two different structures. For this purpose, we calculate the $1-(|S_{11}|^2-|S_{22}|^2)$ for the MS double shielded interposer and the typical GS structure.

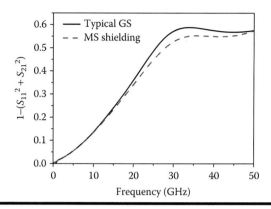

Figure 6.9   Comparison of dielectric and metal losses.

The results are shown in Figure 6.9. As we know, $|S_{11}|^2$ is the reflection power at port 1, and $|S_{21}|^2$ is the power transmitted from port 1 to port 2. For a lossless media and radiationless system, $|S_{11}|^2 + |S_{22}|^2 = 1$.

From Figure 6.9, it can be found that the power loss of the MS interposer is reduced compared with that of the typical GS interposer. The media loss is usually larger than the radiation loss for the silicon interposers; this shows that the dielectric and metal losses of the MS interposer are smaller than that of the typical GS interposer. This conclusion is in agreement with those in Figures 6.7 and 6.8.

Comparison of the dielectric and metal losses between TSV with MS shielding and TSV without MS shielding is simulated, where the RDL traces are removed and ports are defined on the two ends of TSVs. The material characteristic and structure dimension are just the same as those in Figures 6.1 and 6.2 excepting for the removal of the RDLs. As shown in Figure 6.10, the dielectric and metal losses of

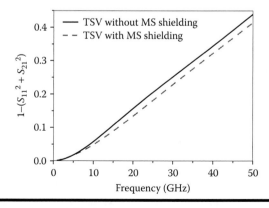

Figure 6.10   Comparison of the dielectric and metal losses between TSV with MS shielding and TSV without MS shielding.

the TSV with MS shielding are reduced compared to that of TSV without shielding in the whole frequency band.

6.1.3 Design Guidelines

In this section, the design guidelines including the highly doped silicon in the surface region and the pattern size of the metallization grids are studied. The parameters are analyzed by the near-end cross talk and transmission (S_{21}) parameter. The structure used for analyzing S_{21} is the same as the model in Figure 6.2. The cross talk model is depicted as in Figure 6.11 with a pitch of 100 μm. Lumped ports are defined between the RDL traces and the meshed metallization layers.

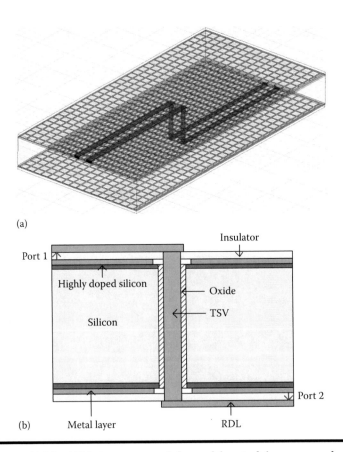

(a)

(b)

Figure 6.11 (a) The MS interposer models used for studying near-end cross talk. (b) Side view.

6.1.3.1 Characteristic Analysis of Highly Doped Silicon Thickness

The highly doped silicon epitaxial layers not only enhance the ohmic contact but also improve the shielding. As the carrier concentration in the highly doped silicon is very high, it acts as a good conductor. Its skin depth is expressed as

$$\delta = \sqrt{\frac{2}{\omega\mu\sigma}} \tag{6.11}$$

where:

   ω is the angular frequency of the electromagnetic field
   μ is the permeability of silicon
   σ is the conductivity of the highly doped layers

Figure 6.12a,b shows the change of S_{21} and cross talk with the thickness of the highly doped layers h. It can be seen that in the low-frequency range, S_{21} of no

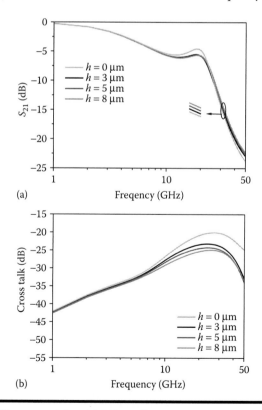

Figure 6.12 **Difference of the shielding effects caused by the thickness of highly doped silicon: (a) S_{21} and (b) near-end cross talk.**

ohmic contact ($h = 0$) is slightly larger than those with ohmic contact. This is because in the low-frequency range, the thickness of the ohmic contact and metallization layer is smaller than the skin depth, and the electromagnetic field can leak into the silicon substrate easily. In the high-frequency range, the skin depth is smaller than h. Thus, the metal grids together with the highly doped layers are effective to prevent electromagnetic field from penetrating into the silicon substrate. As the thickness of the ohmic contact increases, the shielding effect is better.

As depicted in Figure 6.12b, the near-end cross talk is suppressed to an extent. It is because electromagnetic noise in the silicon substrate is bypassed by the low-resistance highly doped region, reducing the coupling between different pairs of TSVs.

6.1.3.2 Characteristic Analysis of Highly Doped Silicon Area

As illustrated in Figure 6.13, "Fully" denotes that the area of the silicon under the metallization grids is fully doped; "Exactly" denotes that the doping region is just under the

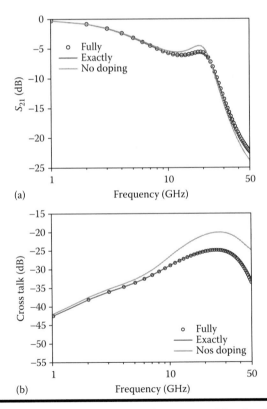

Figure 6.13 Difference of the shielding effects caused by the different areas of highly doped silicon: (a) S_{21} and (b) near-end cross talk.

RDL traces and no doping in other regions; and "No" denotes that no highly doped silicon is grown. It can be seen in the figure that S_{21} parameter and near-end cross talk of Fully and Exactly model is almost the same and has better performance compared with no doping. As a result, in the practical processes, we can just make a shallow highly doped silicon region under the RDL to simplify the fabrication processes.

6.1.3.3 Characteristic Analysis of a Meshed Pattern

Different mesh sizes of the metallization layer are used to analyze their performance, and the results are shown in Figure 6.14. The meshed patterns are in the sizes of $L_w/L_p = 10\ \mu m/50\ \mu m$, $L_w/L_p = 10\ \mu m/26\ \mu m$, $L_w/L_p = 20\ \mu m/50\ \mu m$, and $L_w/L_p = 20\ \mu m/26\ \mu$, respectively. It can be seen that the S_{21} parameter and near-end cross talk of the four different sizes have small distinctions. It is because the size of the grids is much smaller than the wavelength of interest. As a result, the size of the meshed pattern has few impacts on the characteristic of the MS shielding

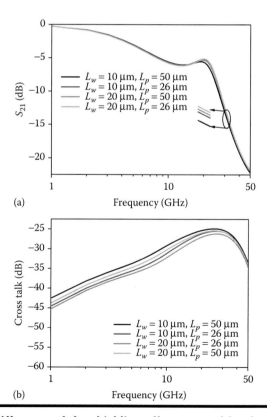

Figure 6.14 **Difference of the shielding effects caused by the size of meshed pattern: (a) S_{21} and (b) near-end cross talk.**

interposer. In the practical semiconductor process, to prevent the delamination of the silicon and metallization layer, we can select a relatively sparse meshing pattern.

6.1.4 Conclusion

In order to improve the high-speed and high-frequency signal propagation characteristics, we propose a double-shielded TSV interposer with the metallization layer directly contacting with the low-resistivity silicon substrate. Shallow highly doped silicon is used to enhance the shielding effects. By using the imaging method, we derive the equivalent circuit model for the proposed structure. The efficiency and accuracy of the equivalent circuit model is validated by the full-wave method. The advantages of the proposed double-shielded interposer with RDL are validated by *S* parameters, compared with the typical GS structure without shielding. It is verified that the proposed structure can provide much better shielding and better confinement of the field around the TSVs. This leads to less signal insertion loss and better suppression of cross talk between adjacent TSVs. At the same time, we propose some useful guidelines to improve the performance of the MS shielded interposer and simplify the processes. The application of low-resistivity silicon reduces the packaging cost. Although the proposed structure needs additional two metal layers, doping, and barrier process, these fabrication processes are easy to be made.

6.2 Compact Integrated Waveguide

As the core technology of 3D integrated circuit, through-substrate via instead of wire bonding is gaining tremendous attraction for its shorter interconnection length, reduced RC delay and parasitic effect, lower power consumption, and flexibility to meet the design requirements.

Besides their use for interconnects, through-substrate vias also can be used for package-level substrate integrated waveguide. Combining the advantage of planar technology with low loss characteristics intrinsic to the nonplanar rectangular waveguide, substrate integrated waveguide technology allows the design of compact lightweight and low-cost devices fully integrated into the substrate [19]. Considerable effort has been devoted to the design and development of PCB-level substrate integrated waveguide in the past few years [20]. However, for the package-level substrate integrated waveguide, due to the low resistivity of silicon substrate ($\rho = 10\ \Omega$ cm), the waveguide structure integrated in silicon interposer always suffers larger transmission loss [19]. New materials and new waveguide structures are required to solve this problem.

Recently, a couple of glass companies have reported large, thin, and low-cost glass wafers with high quality and their usage for TGV [21]. Compared with silicon interposer, especially the expensive and complicated TSV fabrication processes (Bosch process) [22], glass interposer and TGV have a lot of advantages including

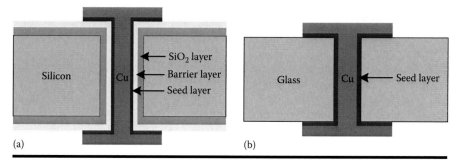

Figure 6.15 **Interposer structures with TSV and TGV technology: (a) silicon interposer with TSV technology and (b) glass interposer with TGV technology.**

high electrical insulation, high optical transparency, low warping and resistance to corrosion, ultra-flat surface, low material cost, and relatively simple TGV production processes [23]. Therefore, glass with TGV technology is a great potential alternative material of silicon to manufacture the interposer for higher frequency 3D system integration applications.

Figure 6.15 shows the schematic view of TSV and TGV structures. TGV interconnection designed in glass interposer, instead of TSV in silicon interposer, eliminates the need for barrier and oxide coating layers before copper-filled processing, which results in reducing via capacitance between copper and interposer tremendously and lowering the electromagnetic interference (EMI) among vias and active and passive circuits. Also, this elimination provides the benefit of cost and complexity reduction. There even presents a new fabrication method of void-free copper-filled TGV for wafer-level RF micro-electro-mechanical system (MEMS) packaging using glass reflow while without seed layer electroplating process in silicon interposer [24] to comprehensively utilize their advantages.

In this section, compact waveguide structures that can be fabricated in glass interposer based on TGV technology instead of silicon interposer are designed so as to decrease the loss mainly caused by the low resistivity characteristic of silicon. Two kinds of waveguide structures are proposed: the transverse electro (TE)-mode waveguide, which is formed by using copper-filled TGV on one edge and air filled TGV on the other edge, and the transverse electromagnetic (TEM) mode waveguide, which is formed by using air-filled TGV on both edges. A full-wave method is applied to analyze the propagation characteristics and electric field distribution of these structures up to 200 GHz. Results demonstrate that the compact TE-mode waveguide has the similar performance to the traditional fully integrated waveguide (where the copper-filled vias are used on both edges), while cutting down the waveguide spacing size almost by 50%. The TEM-mode waveguide shows a low EMI than the traditional planar waveguide structures. The results also show that drilling an array of vias without filling copper along the edge of these compact waveguide structures can concentrate the electromagnetic field inside the waveguides, which leads to less EMI consequently.

6.2.1 TE-Mode Waveguide

6.2.1.1 Compact Waveguide Structures

A compact TE-mode waveguide is proposed and its performance is compared with the full TE-mode waveguide structure. First, a full waveguide structure with planar waveguide transition is illustrated in Figure 6.16, by means of drilling via arrays fully filled with copper. These vias serve as the perfect electric conductor (PEC) as in the rectangular metal waveguide. This waveguide is integrated in a glass interposer with the relative dielectric constant $\varepsilon_r = 6.7$ and the loss tangent $\tan\delta = 0.006$. Each TGV along the waveguide sidewalls has a diameter of $d = 30$ μm and a center-to-center spacing of $s = 40$ μm between two TGVs. Other dimensions of the waveguide include its height $H = 50$ μm, external width $W_{SIW} = 600$ μm and internal width $W = W_{SIW} - 2d$, width $W_{tap} = W/2$ and $W_p = W_{tap}/2$, length $L_{SIW} = 990$ μm, and $L_{tap} = 256.5$ μm and $L_p = L_{tap}/2$.

This full waveguide structure can be equivalent to a glass-filled rectangular waveguide with its metallic sidewalls replaced by arrays of TGVs sufficiently close to each other. Due to the discrete TGV walls, vertical currents can flow through them, whereas longitudinal surface currents of TM modes are unable to propagate into this waveguide. Therefore, it only supports the vertically directed currents of TE waveguide modes and presents a broadband feature for the discrete conducting vias along the sidewalls, whereas it does not allow the propagation of TM modes [25]. In this subsection, we just consider the transmission characteristics of the fundamental TE_{10} mode. According to the rectangular waveguide theory, the cutoff frequency of the TE_{10} mode is

$$f = \frac{1}{2W\sqrt{\varepsilon\mu}} = 107.3 \text{ GHz} \tag{6.12}$$

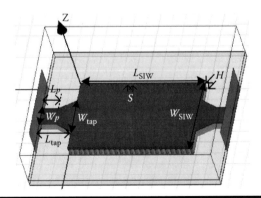

Figure 6.16 3D view of full waveguide structure.

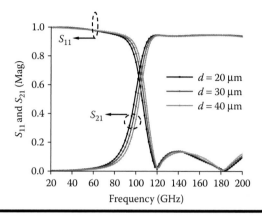

Figure 6.17 **The *S* parameter with the variation of TGV diameter.**

According to Figure 6.17, we can see that increasing TGV diameter will decrease the waveguide internal width *W* leading to the cutoff frequency shifting a little toward higher frequency based on Equation 6.12. However, if we keep the center spacing $s = 40$ µm between TGVs unchanged, the magnitudes of S_{11} and S_{21} parameters will not change too much with the variation of TGV diameter. Therefore, it is unnecessary to pursuit small-size TGV processing, and it makes this waveguide structure more applicable.

The above full waveguide is expensive because it covers a little larger area of the interposer. If we reduce the full waveguide dimension, the cutoff frequency of its fundamental mode will increase, which will eliminate the propagation of low-frequency signals. To solve this problem, we propose the compact half waveguide structures as shown in Figure 6.18, where the width of the compact waveguide is one half of the width of their corresponding full waveguide. In Figure 6.18a, the Half-air waveguide structure is integrated along the edge of interposer, and in Figure 6.18b, a Half-air via waveguide structure is obtained by drilling an array of

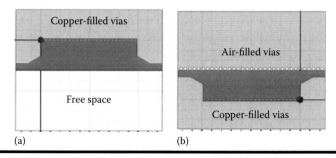

Figure 6.18 **Proposed compact TE-mode waveguide structures: (a) Half-air waveguide and (b) Half-air via waveguide.**

free vias without filling copper along one edge of the half waveguide and an array of copper-filled vias along another edge. For both of these half waveguides, one edge of which can be taken as the PEC and another edge can be taken as the perfect magnetic conductor. Therefore, they can support the same TE_{10} mode as that of their corresponding full waveguide in Figure 6.16.

6.2.1.2 Propagation Characteristics and Electric Field Distribution

Figures 6.19 and 6.20 present the full-wave simulation results for electric field distribution, reflection, and propagation characteristics of the full waveguide and proposed compact half waveguides, respectively.

The *S* parameters and electric field distribution results demonstrate that these two compact half waveguide structures have similar performance to their corresponding full waveguide structures. However, they only occupy 50% area of the full waveguide. Due to the fringing effect (which can be seen from Figure 6.19), part of electric field leaks into the glass or air medium; this results in that the cutoff frequencies of these compact waveguides all shift toward lower frequency.

The electric field leaking into the medium outside of the waveguide will bring about a substantial offset of the cutoff frequency. By comparing the *S*-parameter curves in Figure 6.20 and the electric field distribution in Figure 6.19 of Half-air waveguide structure and Half-air via waveguide structure, it shows free vias filled with air located along the edge of the Half-air via waveguide structure can concentrate the electric field within the waveguide structure and decrease the electromagnetic radiation as a result.

6.2.2 Quasi-TEM-Mode Waveguide

Because the TE-mode waveguide structures proposed in Section 6.2.1 all have cutoff frequencies, they cannot support the propagation of DC and lower frequency signals. In this section, we propose another quasi-TEM-mode compact waveguide as shown in Figure 6.21b, where the planar waveguide is employed and the air-filled vias are placed along its both edges. These air-filled vias serve as the perfect magnetic conductor (PMC) boundary as they do in Figure 6.18b; therefore, the proposed waveguide can support the quasi-TEM mode and hence DC and lower frequency signal propagation along it.

Figure 6.22 plots the electric and magnetic field distribution on the cross section of the proposed planar waveguide with air vias. For both electric field and magnetic fields, their transverse components are quite larger than their longitudinal components. This verifies that such waveguide works at the quasi-TEM mode.

Figure 6.23a,b shows the electric field propagation along the planar waveguides without and with air-filled vias, respectively, where the traditional

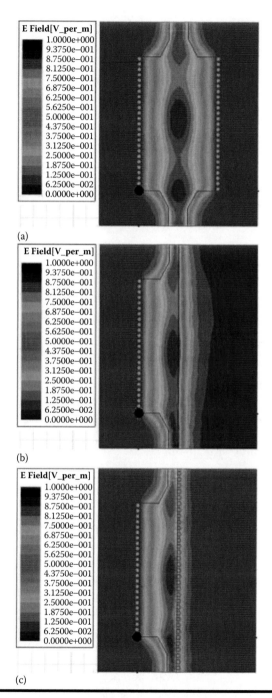

Figure 6.19 **Electric field distribution results at 140 GHz for (a) full waveguide structure, (b) Half-air waveguide, and (c) Half-air via waveguide structure.**

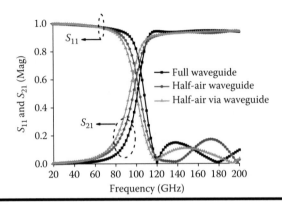

Figure 6.20   Reflection and propagation characteristics.

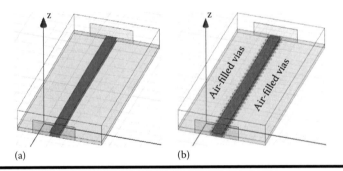

Figure 6.21   (a) Traditional planar waveguide without air-filled vias and (b) planar waveguide with air-filled vias.

planar waveguide without air vias has the same dimension as that of the proposed planar waveguide with air vias. From Figure 6.23, we can see that the electric field of the planar waveguide with air vias is more concentrated within the waveguide region compared with that of the planar waveguide without air vias. This verifies again that the air-filled vias can be used as the PMC boundary to concentrate the electromagnetic fields inside the waveguide.

In Figure 6.24, we study and compare the electromagnetic radiation losses inside the glass interposer. For this purpose, we calculate $1-(|S_{11}|^2+|S_{22}|^2)$ of these two planar waveguides. As we all know, $|S_{11}|^2+|S_{22}|^2 = 1$ is always true for a lossless medium and nonradiative system. To study their EMI property, the dielectric loss of the glass substrate is ignored during the simulation. Therefore, the value of expression $1-(|S_{11}|^2+|S_{22}|^2)$ presents the radiation loss. The results shown in Figure 6.24 demonstrate that the electromagnetic radiation of planar waveguide with air vias dilled along the both edges is smaller than that of planar waveguide without air vias, which results in lowering EMI especially at high frequency.

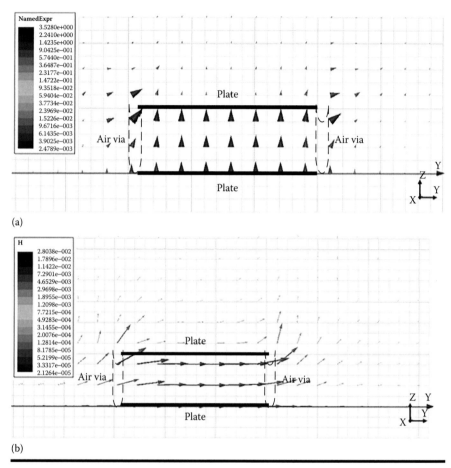

(a)

(b)

Figure 6.22 **(a) Electric and (b) magnetic field vector distribution within planar waveguide with air vias at 100 GHz.**

6.2.3 Conclusion

In this section, we proposed two kinds of compact waveguide structures that can be integrated with glass interposer. Compared with silicon interposer waveguides, they have lower insertion loss. The transmission characteristics represented by S parameters and electric field distribution results of the proposed compact TE-mode waveguide obtained by a 3D full-wave simulator indicate that they behave similarly to their equivalent full waveguides. In addition, they cut down the occupied spacing size almost by 50% and make them more appropriate for high-density 3D integrated circuit system. It also shows that air-filled TGVs can make the electromagnetic field more concentrated within the waveguide structure leading to lower EMI. Meanwhile, the fabrication of air-filled TGVs is easy by using the available TGV process.

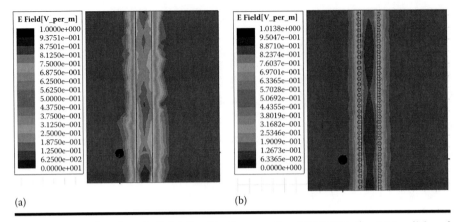

Figure 6.23 The electric field distribution results at 100 GHz for (a) traditional planar waveguide without air-filled vias and (b) planar waveguide with air-filled vias.

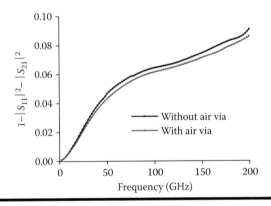

Figure 6.24 Comparison of electromagnetic radiation losses.

References

1. J. Kim, J. S. Pak, J. Cho, et al., High-frequency scalable electrical model and analysis of a through silicon via (TSV), *IEEE Trans. Comp. Packag. Manuf. Technol.*, 1(2), 181–195, 2011.

2. A. E. Engin and S. R. Narasimhan, Modeling of crosstalk in through silicon vias, *IEEE Trans. Electromagn. Compat.*, 55(1), 149–158, 2013.

3. E. P. Li, X. C. Wei, A. C. Cangellaris, et al., Progress review of electromagnetic compatibility analysis technologies for packages, printed circuit boards, and novel interconnects, *IEEE Trans. Electromagn. Compat.*, 52(2), 248–265, 2010.

4. Y. P. R. Lamy, K. B. Jinesh, F. Roozeboom, et al., RF characterization and analytical modelling of through silicon vias and coplanar waveguides for 3D integration, *IEEE Trans. Adv. Packag.*, 33(4), 1072–1079, 2010.

5. B. Banijamali, S. Ramalingam, K. Nagarajanm, et al., Advanced reliability study of TSV interposers and interconnects for the 28 nm technology FPGA, *Electronic Components and Technology Conference*, Lake Buena Vista, FL, 2011, pp. 285–290.
6. N. Tanaka, Y. Yoshimura, M. Kawashita, et al., Through-silicon via interconnection for 3D integration using room-temperature bonding, *IEEE Trans. Adv. Packag.*, 32(4), 746–753, 2009.
7. C. S. Selvanayagam, J. H. Lau, X. W. Zhang, et al., Nonlinear thermal stress/strain analyses of copper filled TSV (through silicon via) and their flip-chip microbumps, *IEEE Trans. Adv. Packag.*, 32(4), 720–728, 2009.
8. J. Li, X. C. Wei, X. J. Wang, et al., Double-shielded interposer with highly doped layers for high-speed signal propagation, *IEEE Trans. Electromagn. Compat.*, 56(5), 1210–1217, 2014.
9. X. C. Wei, X. J. Wang, D. C. Yang, et al., Design of compact and low-EMI waveguide structures based on through glass vias, *Electromagnetic Compatibility*, Tokyo, Japan, May 12–16, 2014.
10. J. Kim, E. Song, J. Cho, et al., Through silicon via (TSV) equalizer, *IEEE Electrical Performance of Electronic Packaging and Systems*, 2009, pp. 13–16.
11. J. J. Tang, X. Chen, G. W. Xu, et al., A novel wafer-level metal/BCB interconnection between both sides of wafer using TSV and its microwave performance, *Electronic Components and Technology Conference*, San Diego, CA, May 29–June 1, 2012, pp. 2121–2128.
12. S. W. Ho, S. W. Yoon, Q. E. Zhou, et al., High RF performance TSV silicon carrier for high frequency application, *Electronic Components and Technology Conference*, Orlando, Florida, December 9–12, 2008, pp. 1946–1952.
13. J. Buechler, E. Kasper, P. Russer, et al., Silicon high-resistivity-substrate millimeter-wave technology, *IEEE Trans. Electron Devices*, 33(12), 2047–2052, 1986.
14. R. Y. Yang, C. Y. Hung, Y. K. Su, et al., Loss characteristics of silicon substrate with different resistivities, *Micro. Opt. Technol Lett.*, 48(9), 1773–1776, 2006.
15. D. C. Yang and X. C. Wei, Impedance calculation of power and ground planes by using imaging methods, *Asia-Pacific Symposium Electromagnetic Compatibility*, Singapore, May 21–24, 2012, pp. 37–40.
16. J. S. Pak, J. Kim, J. Cho, et al., On-chip PDN design effects on 3D stacked on-chip PDN impedance based on TSV interconnection, *Electrical Design of Advanced Package & Systems Symposium*, Singapore, December 7–9, 2010, pp. 1–4.
17. Y. Wang, F. Cao, M. H. Ding, et al., Diffusion barrier performance of Zr–N/Zr bilayered film in Cu/Si contact system, *Microelectron Reliab.*, 48(11–12), 1800–1803, 2008.
18. Y. Wang, F. Cao, X. D. Yang, et al., Non-crystalline Zr–Si diffusion barrier for Cu/Si contact system under different sputtering power, *J. Non-Crystalline Solids.*, 355(52–54), 2567–2570, 2009.
19. F. Fesharaki, C. Akyel, and K. Wu, Broadband substrate integrated waveguide edge-guided mode isolator, *Electron. Lett.*, 49(4), 269–271, 2013.
20. D. Deslandes and K. Wu, Accurate modeling, wave mechanisms, and design considerations of a substrate integrated waveguide, *IEEE Trans. Microwave Theory Tech.*, 54(6), 2516–2525, 2006.
21. C. Kim and Y. K. Yoon, High frequency characterization and analytical modeling of through glass via (TGV) for 3D thin-film interposer and MEMS packaging, *IEEE 63rd Electronic Components and Technology Conference*, Las Vegas, Nevada, May 28–31, 2013, pp. 1385–1391.

22. V. Sukumaran, T. Bandyopadhyay, Q. Chen, et al., Design, fabrication and characterization of low-cost glass interposers with fin-pitch through-package-vias, *IEEE 61st Electronic Components and Technology Conference*, Lake Buena Vista, Florida, May 31–June 3, 2011, pp. 583–588.
23. Y. Sun, D. Q. Yu, R. He, et al., The development of low cost through glass via (TGV) interposer using additive method for via filling, *IEEE 13th International Conference on Electronic Packaging Technology & High Density Packaging*, Guilin, Guangxi, China, August 13–16, 2012, pp. 49–51.
24. J. Yong, S. W. Lee, S. K. Lee, et al., Wafer level packaging for RF MEMS devices using void free copper filled through glass via, *IEEE 26th International Conference on Micro Electro Mechanical Systems*, Taipei, Taiwan, January 20–24, 2013, pp. 773–776.
25. F. Xu and K. Wu, Guided-wave and leakage characteristics of substrate integrated waveguide, *IEEE Trans. Microw. Theory Tech.*, 53(1), 66–73, 2005.

Chapter 7

New Structures and Materials

Due to the more and more rigorous electromagnetic compatibility (EMC) standards and regulations for the low emission and high immunity, traditional EMC control and design methods face the great challenge. Currently, new developments in materials and periodic structures provide potential solutions to this problem. In this chapter, we will talk about the applications of high-impedance surface (HIS) and graphene films on EMC designs.

HIS has two unique characteristics: the surface-wave suppression and in-phase reflection properties. It was proposed and employed in microwave and antenna engineering. We use it to eliminate the surface-wave noise propagation. First, the effect of the metal shielding box on the stability of radio frequency (RF) chips is modeled by a field–circuit hybrid method, and then the HIS is designed to improve the stability of the chip. Measurements are carried out for a commercial amplifier module with its shielding box. Good correlations between measurement results and simulation results are obtained, which verify the HIS performance. We also show the application of HIS on the performance enhancement of patch antennas working at Wi-Fi frequency (2.45 GHz) [1]. The input impedances and radiation patterns of the traditional patch antenna and the HIS patch antennas are compared through both the full-wave simulation and the measurement. These results show that the HIS can suppress the surface waves and reduce back lobes. Meanwhile, the effect of numbers of HIS cells on the antenna performance is studied.

As a two-dimensional (2D) material, graphene film provides a potential solution to the complex EMC problems. In this chapter, we demonstrate that it can be used as a thin absorbing film. High-quality single-layer graphene grown by chemical vapor deposition (CVD) method has a sheet impedance ranging from 300 to 600 Ω. A coaxial testing method is proposed for the electromagnetic characterization of the graphene film [2]. After that, an efficient and accurate transmission line (TL) model is proposed for the design of such transparent graphene film absorber at 38.5 GHz. The result of the TL model is in good agreement with that of the three-dimensional (3D) full-wave simulation. Finally, the sample of the graphene absorber is fabricated, and a testing method is proposed to characterize the absorbing property of the absorber sample.

7.1 High-Impedance Surface

With the wide and in-depth application of the electromagnetic wave in engineering, there is a great demand of the new materials, which can be used by engineers to control the electromagnetic wave in a more flexible way. As the artificial and periodic structures, metamaterials and metasurfaces [3] are proposed for this purpose, which show interesting and abnormal electromagnetic properties compared with the traditional metal or dielectric materials.

HIS can be taken as one kind of metasurface. One common way to fabricate a HIS is to etch the metal layers of a printed circuit board (PCB) into cells to form the periodic structures. At the resonant frequency of those cells, the surface impedance of the HIS becomes very high, so that it can eliminate the surface-wave propagation along its surface [4]. At the same time, there are no currents flowing on the surface of HIS at the resonant frequency either, which makes the HIS like a perfect magnetic conductor (PMC) [5]. Therefore, the HIS is also referred to as the artificial magnetic conductor in some literature [6]. When an electromagnetic wave illuminates on the HIS, the reflected wave will be in-phase with the incident wave.

These two unique characteristics, the surface-wave suppression and the in-phase reflection, make the HIS find many potential applications in the electromagnetic engineering. It is used as the power–ground plane for the simultaneous switching noise mitigation in high-speed circuits [7]. It had been employed to enhance the gain of the antennas and decrease the backward radiation [8]. A TL model is proposed in [9] for the simple and accurate analysis of the HIS. Although it shows potential benefits, there is still challenge for HIS to be used in the practical engineering. One difficult is its cell's larger size versus the space constraints in many applications.

In this section, the performance of HIS used in two practical electromagnetic engineering problems is explored, where two kinds of HISs are tested: the mushroom-type HIS used for the back lobe reduction of microstrip patch antennas and the cavity-type HIS used for the noise reduction inside a metal shielding box.

7.1.1 Basic of HIS

The fundamentals of the mushroom-type and cavity-type HISs are described in Sections 7.1.1.1 and 7.1.1.2.

7.1.1.1 Mushroom-Type HIS

The mushroom-type HIS is a commonly used HIS. Its geometry is shown in Figure 7.1a, where the top metal layer of the PCB is etched into many patches, and each patch is connected to the bottom metal ground plane through the via. The gap between adjacent patches forms the capacitance C, whereas the vias and the ground plane form the inductance L. Therefore, the unit cell of the HIS can be equivalent to an LC parallel circuit as shown in Figure 7.1b.

The resonant frequency of the LC parallel circuit is $\omega_0 = 1/\sqrt{LC}$. For $\omega < \omega_0$, the impedance magnitude of L is smaller than that of C, so a large amount of the current flowing on the HIS surface will go through L, which results in that the HIS presents an inductive surface. For this case, the HIS can support transverse magnetic (TM) wave propagation along its surface. However, when $\omega > \omega_0$, the HIS presents a capacitive surface and can support transvers electric (TE) wave propagation along its surface. When $\omega = \omega_0$, the impedance of the HIS approaches infinity, which will stop the current flowing along its surface, and also eliminate the surface-wave propagation along its surface.

7.1.1.2 Cavity-Type HIS

Besides the lumped elements LC, the patch cavity resonance also can be used to form the HIS. As shown in Figure 7.2a, the cavity-type HIS is similar to the mushroom-type HIS except without the shorting vias. The surface current of the HIS is also shown in Figure 7.2a; it flows on the patch and the ground plane. The discontinuity between the surface currents on the patch and the ground plane is compensated by a vertical displacement current flowing on the edge of the patch. This displacement current goes through the edge impedance Z (Z is for looking into the patch) of the patch.

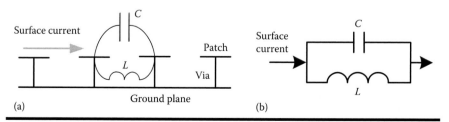

Figure 7.1   **(a) The mushroom-type HIS and (b) its equivalent circuit.**

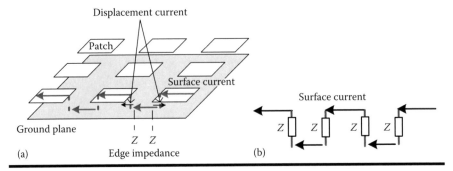

Figure 7.2 **(a) The cavity-type HIS and (b) its equivalent circuit.**

The equivalent circuit to describe the flowing of the surface current is plotted in Figure 7.2b. Because the gap between patches is large enough, the parasitic capacitance of the gap is very small and not considered in the equivalent circuit. Because the distance between the patch and the ground plane is much smaller than the wavelength of interest, each patch together with the ground plane can be taken as a cavity with the perfect electric conductor (PEC) on the top and bottom and the PMC along the edges of the patch. At the resonant frequency of the cavity, the edge impedance Z approaches to infinity, which will greatly suppress the surface current and then the surface-wave propagation along the HIS.

7.1.2 Applications of HIS

7.1.2.1 Shielding Box

Due to the ever-increasing working frequency and integration density, the electromagnetic interference between RF chips becomes much serious than before. In order to reduce the noise coupling between the noisy amplifier and the sensitive circuits, metal shielding boxes are widely employed inside modern electronic products. However, the shielding box introduces an unwanted electromagnetic feedback from the output port of the power chip to its input port, and finally changes the total amplifier gain [10]. Such gain variation is uncontrollable, especially considering that the designers of the power chip and the shielding box are not the same engineer.

The accurate and efficient modeling of the effect of the shielding box on the amplifier is a challenge for available solvers. Analysis formula based on the TL theory is proposed to calculate the shielding effectiveness of the shielding box with apertures [11,12]. This formula is efficient but limited to the shielding box with a rectangular shape. However, full-wave methods can be employed for the shielding box with arbitrary shapes, such as the finite element method and the finite-difference time-domain method. However, those methods require a long computing time and large memory consumption. In order to release the computing burden, the method of moment with spatial Green's function [13,14] can be employed. Unfortunately, their applications are limited to the closed boxes with regular shapes.

Besides the modeling of the enclosed power chip, how to design the shielding box to reduce its effect on the amplifier gain is also an important problem. The absorbing material can be attached to the surface of shielding box to reduce the coupling between the shielding box and the power chip. However, the experiment shows that it can reduce the efficiency of the enclosed power chip. HISs are artificial materials with periodic structures, which have interesting applications in circuit designs. HIS provides a potential solution for the shielding box design.

In the following, the effect of the shielding box on the enclosed power chip is modeled by using the field–circuit hybrid method, where the coupling between the shielding box and the chip is extracted and modeled as a feedback S parameter. Based on the proposed field–circuit hybrid method, a slot-type HIS is designed to reduce the unwanted feedback S parameter.

7.1.2.1.1 Field–Circuit Hybrid Modeling

A commercial amplifier module used in the communication is employed for the demonstration. This amplifier module includes the power chip and its input and output impedance matching components. All of them are enclosed in a metal shielding box with an irregular shape. The chip has a narrow working band around 2.1 GHz. The shielding box is designed so that its resonant frequencies do not cover the working band of the chip.

Measurement results in Figure 7.11 show that after enclosed by the shielding box, the gain (G_{PEC}) becomes larger than that without the shielding box ($G_{\mathrm{Free\ Space}}$). The presence of the shield box modifies substantially the electromagnetic environment surrounding the power chip. At high frequency, the internal surfaces of the shielding box will introduce an electromagnetic feedback from the output port to the input port, and this feedback together with the original amplifier gain forms a closed loop, so that the total gain with the shielding box is different from the nominal value of the gain. This feedback is modeled by using the following field–circuit hybrid method.

For the well-designed power chip and its impedance matching components, reflections at their terminals are much small. Therefore, a simplified signal flow graph as shown in Figure 7.3 is used to calculate the total gain of the amplifier as

$$G \approx \frac{G_0}{\left|1 - S_{25}S_{43}\right|} \tag{7.1}$$

$$G_0 = \left|S_{21}S_{43}S_{65}\right| \tag{7.2}$$

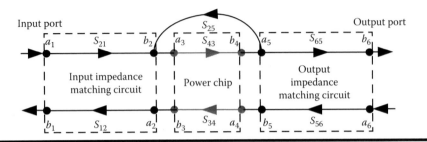

Figure 7.3 Simplified signal flow graph for the amplifier module.

where S_{43}, S_{21}, S_{65}, and S_{25} are the nominal gains of the power chip, transmission coefficients of the input and output impedance matching circuits, and the unwanted feedback coefficient, respectively. S_{21}, S_{65}, and S_{25} are extracted from the full-wave simulation of the input and output impedance matching circuits and the shielding box. S_{43} is obtained from the datasheet of the power chip.

The accuracy of the proposed field–circuit hybrid method is verified by comparing its simulation results with the measurement results, as shown in Figure 7.4. The measurement setup is shown in Figure 7.5. Here, different boundary conditions are applied on the shielding box: the free-space condition where the shielding box is removed and the electric perfect conductor condition with the presence of the shielding box. These two boundaries provide different amounts of the electromagnetic coupling in the air region above the power chip, which result in different S_{25}. Good correlation between results from the simplified signal flow graph and the measurement can be observed.

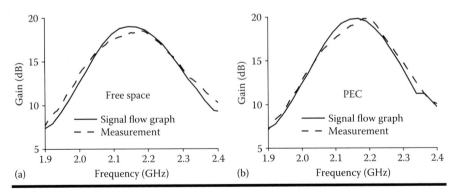

Figure 7.4 Total gains obtained by using the signal flow graph and from the measurement: (a) free-space (without the shielding box) and (b) PEC (with the presence of the shielding box) boundaries.

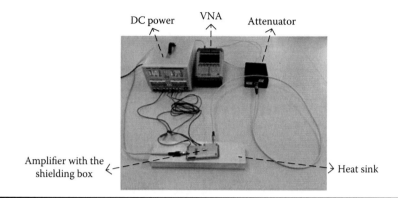

Figure 7.5   Setup for the amplifier gain measurement.

Next, the effects of different boundaries on $|S_{25}|$, $|S_{21}|$, and $|S_{65}|$ are compared in Figure 7.6, where the magnetic perfect conductor boundary means the up surface of the shielding box is set as the PMC boundary. The PMC boundary can be taken as the ideal case of the HIS, so its effect on these S parameters is compared with those of PEC and free-space boundaries. We can see that the transmission coefficients of the input and output impedance matching circuits ($|S_{21}|$ and $|S_{65}|$) do not change too much for different boundaries, whereas the value of $|S_{25}|$ shows great differences under different boundaries. $|S_{25}|$ represents the unwanted coupling between the input and the output of the power chip, which is introduced by the shielding box. Although its value is smaller, it forms a loop with the large $|S_{43}|$ and becomes the major contribution to the unstable amplifier gain.

Because different boundaries do not have much effect on $|S_{21}|$ and $|S_{65}|$, and also considering $|S_{43}|$ does not change with the environment, we can take G_0 of Equation 7.2 as a reference gain for this amplifier module. G_0 is almost a constant for different boundaries. From Equation 7.1, we can see that the instability of the amplifier gain is due to the change of S_{25} with boundary conditions. S_{25} is produced by the electromagnetic coupling through the air region above the power chip. Different boundaries produce different S_{25}, and then result in different amplifier gains.

Figure 7.6a shows that the PMC boundary (or HIS) gives a smaller $|S_{25}|$ and then a gain more close to G_0. To further explain the benefit of the PMC boundary (or HIS), the change of $1-S_{25}S_{43}$ with the frequency is plotted in Figure 7.7 for three different boundaries, where the abscissa denotes the real part and the vertical axis denotes the imaginary part. In Figure 7.7, a dot circle with the center at the origin and the radius of 1 is also plotted. If $1-S_{25}S_{43}$ falls on this circle, according to Equation 7.1, $G \approx G_0$; if $1-S_{25}S_{43}$ falls inside of this circle, $G > G_0$; if $1-S_{25}S_{43}$ falls outside of this circle, $G < G_0$. Because $|S_{25}|_{\text{PMC}} < |S_{25}|_{\text{Free Space}} < |S_{25}|_{\text{PEC}}$ in Figure 7.6a, the variation range of $1-S_{25}S_{43}$ is increased from PMC, free space, to

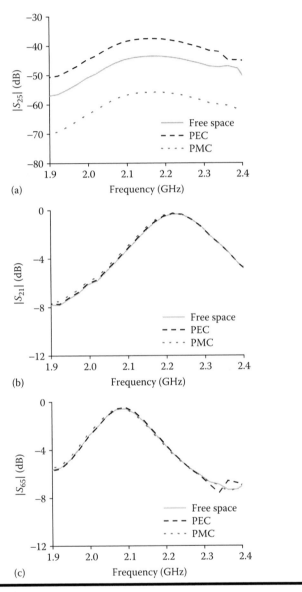

Figure 7.6 **(a) $|S_{25}|$, (b) $|S_{21}|$, and (c) $|S_{65}|$ under three different boundaries.**

PEC, which results in a more instable G. Figure 7.8 shows the comparison between G_0 and total amplifier gains under different boundary conditions. PMC boundary means that the up surface of the shielding box is open circuited, which greatly suppresses the surface-wave propagation along the internal surface of the shielding box; therefore, G_{PMC} is much close to G_0. However, PEC boundary means that the up

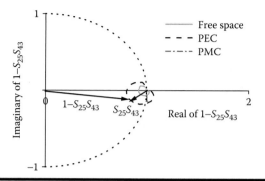

Figure 7.7 **Change of 1–$S_{25}S_{43}$ with the frequencies for different boundaries.**

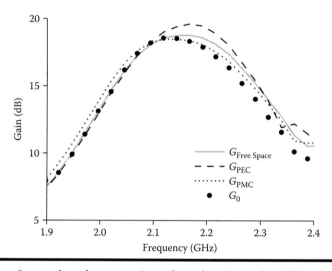

Figure 7.8 **Comparison between G_0 and total gains under different boundary conditions.**

surface of the shielding box is short circuited, so G_{PEC} shows the largest deviation from G_0. $G_{Free\ Space}$ is just in between G_{PMC} and G_{PEC}.

7.1.2.1.2 Measurement Results

The above analysis suggests that PMC boundary can give a gain much close to G_0. However, there is no natural PMC material. The HIS, which is an artificial material, can serve as the PMC material during its working frequency band. We design a slot-type HIS as shown in Figure 7.9a and b. The substrate is FR4 with a dielectric constant of 4.4 and a thickness of 2 mm.

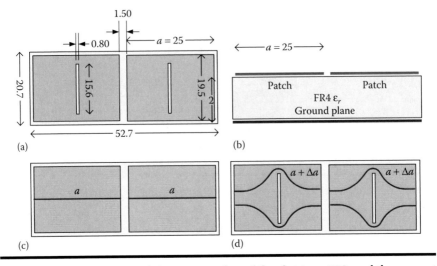

Figure 7.9 **(a) Top view and (b) cross section of the slot-type HIS, and the current path of the principle resonant mode of the cell (c) without the slot and (d) with the slot (unit: mm).**

The cell in Figure 7.9a is a rectangular patch. Because the thickness of the substrate is much smaller than the interesting wavelength, the patch and ground planes form a cavity. At the resonant frequencies of the cavity, the periphery of the patch shows very high impedance. It will suppress the surface-wave propagation along the surface of the substrate. This HIS is attached to the up internal surface of the shielding box and can serve as the PMC boundary.

The principle resonant frequency of the cell is

$$f = \frac{c}{2a\sqrt{\varepsilon_r}}$$
(7.3)

where:
  c is the speed of light in vacuum
  ε_r is the dielectric constant of the substrate
  a is the length of the patch

At the working frequency of the amplifier (2.1 GHz), from Figure 7.3 the size of the patch is too large to be attached to the shielding box. To solve this problem, a slot is etched on the patch as shown in Figure 7.9a. This slot increases the path of the current of the principle resonant mode, as shown in Figure 7.9c and d, so that the principle resonant frequency is reduced, or in another word, the patch size can be reduced and the whole HIS can be attached to the shielding box.

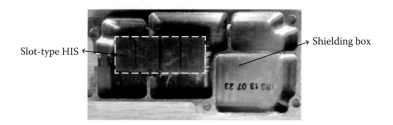

Figure 7.10   The fabricated slot-type HIS, which is attached to the shielding box.

Figure 7.10 shows the fabricated slot-type HIS, which is attached on the up internal surface of the shielding box by the conductive adhesive. The amplifier gain with the HIS-attached shielding box G_{HIS} is measured by using the ZVH8 vector network analyzer (VNA) (from Rohde and Schwarz company, German).

Figure 7.11 plots the measured amplifier gains with the original shielding box G_{PEC}, with the HIS-attached shielding box G_{HIS} and without the shielding box $G_{Free\ Space}$. From this figure, we can see that after attaching the HIS, the amplifier gain is reduced and is more close to the gain without the shielding box, especially during the working frequency band around 2.1 GHz. This verifies the efficiency of the proposed slot-type HIS. It should be noted that due to the dielectric and metal losses, the surface impedance of the practical HIS is not infinite. Its value is between those values of PEC and PMC boundaries. That means it still has a small $|S_{25}|$, but it is much closer to the $|S_{25}|$ of the free-space boundary than the PEC boundary does.

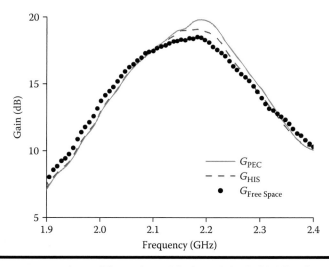

Figure 7.11   Measured amplifier gains with the original shielding box G_{PEC}, with the HIS-attached shielding box G_{HIS}, and without the shielding box $G_{Free\ Space}$.

7.1.2.2 Antenna Design

The surface-wave propagation along the substrate and the diffraction by the edge of the patch antenna result in the antenna's back lobe, which also reduce the antenna gain and introduce unwanted coupling between patch antennas and its nearby circuits. In this section, the three-layer mushroom-type HIS is designed to suppress the surface wave and improve the performance of the patch antenna working at Wi-Fi frequency (2.45 GHz) [1].

7.1.2.2.1 HIS Antenna Design

Three kinds of patch antennas are designed. One is the rectangular patch antenna without HIS (named as "reference antenna" in the following). The other is the rectangular patch antenna with three loops of HIS cells surrounding around the patch (named as "three loops of HIS antenna"). The last one is the rectangular patch antenna with five loops of HIS cells surrounding around the patch (named as "five loops of HIS antenna"). The patch sizes of three antennas are the same, and they have the same FR4 substrate with the thickness of 1.6 mm. They all use the subminiature A (SMA) coaxial feed. The stopband of the three-layers HIS is specially designed to cover the antenna working frequency 2.45 GHz. The side view of the HIS is shown in Figure 7.12. In order to reduce the cell size, the three-layer (two layers of patches and one ground plane) design is used.

The fabricated antennas are shown in Figure 7.13, where their dimensions are shown in Table 7.1.

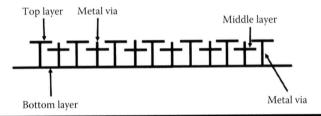

Figure 7.12   Side view of the three-layer HIS.

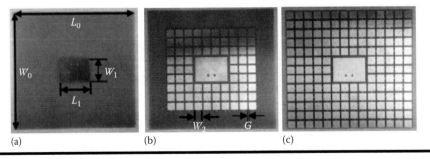

Figure 7.13   (a) The reference antenna, (b) three loops of HIS antenna, and (c) five loops of HIS antenna.

Table 7.1   The Dimensions of Three Kinds of Antennas

Parameter	Values (mm)
L_0	150
W_0	140
L_1	37.26
W_1	28
W_2	9.9
G	1.1

7.1.2.2.2 Measurement Results

We measure the radiation patterns of three kinds of antennas in the anechoic chamber by using standard horn antennas. The measurement setup is shown in Figure 7.14. The radiation pattern in the E-plane and H-plane of three kinds of antennas are shown in Figure 7.15.

The measured gain with HIS is about 0.5 dB larger than that of the reference antenna. We can make a conclusion from the radiation pattern in the E-plane and H-plane: HIS can effectively suppress the back lobe radiation and then enhance the antenna maximum directivity and gain.

Another interesting thing is that the back lobe of the three-loop HIS antenna is about 1 dB less than that of the five-loop HIS antenna in the E-plane, and about 2 dB less than that of the five-loop HIS antenna in the H-plane. The HIS itself will also introduce small electromagnetic wave reflection, which will propagate to

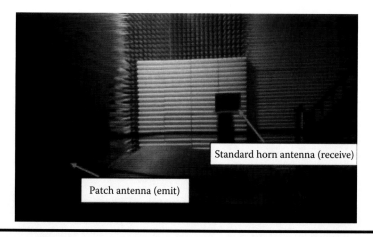

Figure 7.14   Measurement setup in the anechoic chamber.

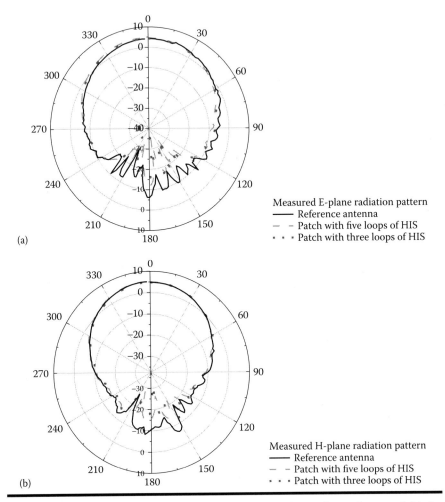

Figure 7.15 **The radiation pattern of three antennas in (a) the E-plane and (b) the H-plane.**

the substrate edge and contribute to the back lobe due to the edge diffraction. The interaction between the original surface wave and the reflected waves from the HIS could be complex, so it does not mean that the more cells HIS has, the better performance antenna achieves.

7.1.3 Conclusion

In this section, from the view of the practical engineering, the performance of the HIS is verified. We design a mushroom-type HIS to reduce the back lobe of the microstrip patch antenna and a cavity-type HIS to reduce the unwanted coupling inside the metal shielding box. Measurement result shows that the designed HIS can greatly improve the performance of the antenna and shielding box. The results in this section also show that the nearby objects could have their effect on the elimination of the surface waves by using the HIS, because the HIS is an artificial material rather than a real material. Therefore, to achieve the best performance of the HIS, the codesign of the HIS and its nearby objects is required.

7.2 Graphene

Graphene is a 2D material that has been a hot topic since its discovery. Beneficial from its outstanding electrical and mechanical properties, tremendous efforts have been made to apply this material into electromagnetic components such as antennas [15,16], frequency selective surfaces [17], and phase shifters [18]. However, most work pays attention to its application in terahertz band, which is mainly because graphene has a remarkable dynamic inductance in terahertz band, and its sheet impedance can be tuned by the external electric or magnetic field [19], which results in many interesting applications such as terahertz plasmonic antennas, terahertz modulators, and tunable metamaterials. In the microwave band, the sheet resistance of graphene still can be tuned by the electric or magnetic bias, but its sheet inductance is negligible, which does not support the propagation of surface plasmon wave and limits its applications in most microwave components.

However, for most of electromagnetic applications, gigahertz band or microwave band is the major interesting frequency band. In this section, we will explore the potential applications of the 2D materials for gigahertz band EMC engineering.

7.2.1 Electromagnetic Characterization

To date, various synthesis techniques of graphene are proposed, such as mechanical exfoliation, CVD, and SiC epitaxial growth. The property of graphene synthesized from different techniques varies tremendously. The electrical performance of graphene-based device is strongly dependent on the properties of graphene layer itself, so that the characterization of graphene without deteriorating is of ultimate importance. Developing a rapid, simple, and reliable technique to characterize the basic electronic properties of graphene is emergently required.

To determine the scattering rate of graphene at microwave frequency is usually through I–V measurement, which will perturb or even deteriorate the intrinsic properties of graphene under test [20,21]. A coplanar waveguide (CPW) structure, which can guide quasi-transverse electromagnetic mode waves, is usually utilized to measure the surface impedance of graphene [22,23]. However, the CPW-based characterization methods need a de-embedding procedure to remove the high contact impedance between the graphene and the metal [24], which introduces an inevitable postprocessing uncertainty. Recently, a noncontact method by using rectangular waveguides with TE_{10}-mode normal incident waves has been proposed to characterize the surface impedance of monolayer graphene (MLG) at micro- and millimeter-wave frequencies, but de-embedding is still necessary to eliminate the effect of substrates and air inside the waveguides [25]. Another noncontact method by using high-Q (quality factor) microwave dielectric resonator perturbation technique is proposed [26], whereas its testing frequency band is limited to its narrow resonant frequency band.

In this section, we propose a contact-free characterization technique to obtain the surface conductance and scattering rate of graphene [1], in which the de-embedding procedure is not necessary. Teflon ($\varepsilon_r = 2.0$, loss tan $\delta = 2.8 \times 10^{-4}$) is chosen as the substrate to avoid issues related to the inversion layer occurring at the Si–SiO_2 interface. Three sets of graphene samples were considered: pristine MLG, double-layer graphene (two MLG stacked [t-MLG]), and doped MLG (d-MLG), in order to investigate the impact of different layer properties on the surface conductivity and scattering rate. By combining Raman spectroscopy and microwave measurement with Amphenol Precision Connector (APC-7), from Hangzhou Electronic, China coaxial connectors, we investigate the surface scattering rate of graphene at microwave frequency. The Fermi energy is extracted from Raman spectrum and the surface conductivity of graphene is characterized by the APC-7 coaxial connectors. Because coaxial connectors are used, the proposed measurement method is suitable for a wider frequency band from DC to tens of gigahertz.

7.2.1.1 Fabrication of Graphene Film

MLG is grown on a 50 μm-thick copper foil at 1050°C by using an Aixtron Black Magic® 2″ CVD system, in which methane is chosen as a carbon-containing precursor. To transfer the graphene, we first spin coat 300 nm of polymethyl methacrylate (PMMA) film onto one side of the graphene/Cu foil and strip the graphene on the other side with O_2 plasma. The copper is then etched away using copper etchant ($CuSO_4$:HCl:H_2O) by floating the PMMA-coated foil on the surface of the etchant bath. The PMMA/graphene on copper foil is rinsed by HCl/H_2O (1:10) and deionized (DI) water for three times in order to remove the eventual metal residues on the back side. The cleaned PMMA/graphene thin films, as shown in Figure 7.16a, are picked up from water by white ring-shaped Teflon substrates. Figure 7.16b displays the optical microscope image of a ring-shaped Teflon substrate (white part) covered by MLG after PMMA removal by acetone vapor. The graphene–Teflon stack is then put into an oven at 110°C for one hour to further improve the adhesion force between graphene and Teflon substrate. Figure 7.16c presents the enlarged scanning electron microscope (SEM) image of graphene on Teflon substrate, where the white ring region in high contrast is the Teflon substrate covered by graphene and the central region in low contrast is the suspended graphene.

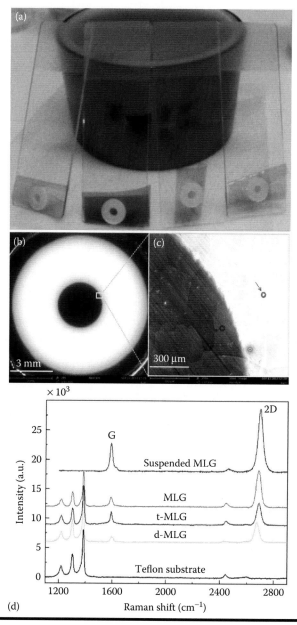

Figure 7.16 **Sample preparation and Raman characterization of MLG/Teflon system: (a) pickup and drying of graphene on Teflon substrates after etching and DI rinse; (b) optical microscope image of graphene on a Teflon substrate after PMMA removal; (c) SEM image for the rectangle zone of (b); (d)** *from bottom to top:* **Raman spectra of the naked Teflon substrate (black), d-MLG on Teflon substrate, t-MLG on Teflon, MLG on Teflon, and the suspended graphene.**

All the samples transferred on Teflon substrates are measured by 532 nm laser Raman spectroscopy depicted in Figure 7.16d. Our MLG samples have clear MLG characteristic peaks with G band at 1586 cm^{-1} (for C=C aromatic ring chain vibration), 2D band at 2695 cm^{-1} (for double resonance intervalley Raman scattering), negligible D band, and $I(G)/I(2D) = 0.29$ with pure Lorentzian function of 2D band, showing its high quality of our CVD graphene growth. The doping density of this sample is estimated as $n = 2.30 \times 10^{12}$ cm^{-2} ($E_f = 0.18$ eV).

t-MLG is prepared in a similar way as MLG by repeating twice the process. The doping effect of HNO$_3$ on graphene was reported in [27]. We prepare d-MLG by putting transferred undoped graphene sample into a beaker containing 65% HNO$_3$ for 10 seconds. Raman spectra of t-MLG and d-MLG are shown in Figure 7.16d as well; the doping densities are estimated as $n = 1.08 \times 10^{13}$ cm^{-2} ($E_f = 0.38$ eV) and $n = 2.32 \times 10^{13}$ cm^{-2} ($E_f = 0.56$ eV) for t-MLG and d-MLG, respectively.

7.2.1.2 Equivalent Circuit of Graphene Film

As for electric characterization in macroscopic scale, when the size of graphene is much larger than the mean free path of the carriers, carrier transport in graphene degrades from ballistic to classical transportation. The surface conductivity σ can be described from Kubo formula [28] as

$$\sigma(\omega, E_f, \Gamma, T) = \frac{je^2(\omega - j\Gamma)}{\pi\hbar^2} \left[\frac{1}{(\omega - j\Gamma)^2} \int_0^\infty \left(\frac{\partial f_d(\varepsilon)}{\partial \varepsilon} - \frac{\partial f_d(-\varepsilon)}{\partial \varepsilon} \right) d\varepsilon \right.$$

$$\left. - \int_0^\infty \frac{f_d(-\varepsilon) - f_d(\varepsilon)}{(\omega - j\Gamma)^2 - 4(\varepsilon/\hbar)^2} d\varepsilon \right]$$

(7.4)

where:

$\hbar$ is the reduced Planck's constant

e is the elementary charge

ω is the angular frequency

$\Gamma = 1/\tau$ is the carrier scattering rate in graphene

τ is the relaxation time

Fermi–Dirac distribution is given by $f_d(\varepsilon) = 1/[1 + \exp[(\varepsilon - E_f)/k_B T]]$ at temperature T ($T = 300$ K is chosen in this work) with Fermi energy of graphene $E_f = \hbar v_f(\pi n)^{1/2}$, where $v_f = 1.1 \times 10^6$ m/s is the Fermi velocity.

The Kubo formula shows that the surface conductivity of graphene has a frequency-independent value at the microwave band, and the imaginary part of surface conductivity is close to zero. Based on this, σ can be extracted from measuring the frequency response

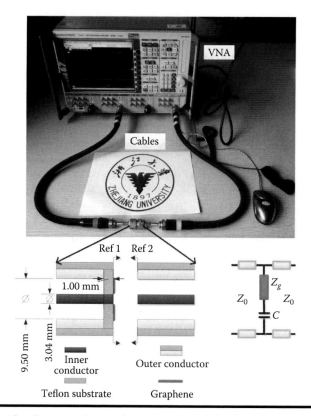

Figure 7.17   The diagram of experiment setup and the lumped equivalent circuit of MLG sandwiched between two APC connectors.

of graphene under transverse electromagnetic (TEM) mode illumination by the proposed technique. By inserting σ and the Fermi energy E_f obtained from Raman spectrum into Equation 7.5, the carrier scattering rate Γ and the relaxation time τ can be deduced.

The diagram of experiment setup is illustrated in Figure 7.17, in which the graphene-covered Teflon substrate is sandwiched between two APC-7 connectors. The APC connectors are connected to a VNA (ZVA67 from Rohde and Schwarz, German) through coaxial cables. The outer diameter of the inner conductor and the inner diameter of the outer conductor of APC connectors are 3.04 and 9.50 mm, respectively. To hold the Teflon/graphene under test, a slot with 1.00 mm depth is etched on the outer conductor of the left APC connecter in Figure 7.17.

Before the measurement, the thru-open-short-match calibration standard is used to calibrate the system. In order to consider the dielectric loss in the substrate and increase the accuracy of measurement, a naked Teflon substrate is placed in the slot. This calibration procedure moves the reference planes to the interface of the two APC connecters, as depicted as "Ref 1" and "Ref 2" in Figure 7.17. Therefore, in the proposed APC measurement method, the de-embedding procedure used in the

CPW-based and rectangular waveguide measurements is avoided. As a consequence, the measurement is greatly simplified. Another benefit of using the APC coaxial connectors is that the electromagnetic wave propagating inside is the TEM mode, which can better model the plane wave in free space than those of using the transverse magnetic/transverse electric modes in circular/rectangular waveguide methods [25], and it is suitable for a wider frequency band from DC to tens of gigahertz.

Because the measuring microwave wavelength is six to seven orders of magnitude larger than the thickness of the graphene samples, the series impedance of graphene can be ignored. Therefore, the graphene sample can be regarded as shunt impedance Z_g. We note that the air gaps between the inner/outer conductors and the graphene should be taken into account for the accurate extraction of the surface conductivity. These air gaps can be represented by a capacitance in series with Z_g.

The equivalent circuit of the target graphene sandwiched between the APC connectors is illustrated in Figure 7.17; therefore, the total impedance can be calculated from the measured S parameters by

$$Z_{total} = \frac{2Z_0 S_{21}}{(1 - S_{11})(1 - S_{22}) - S_{12}S_{21}} \tag{7.5}$$

where:

$Z_{total} = Z_g + 1/j\omega C$ is the total impedance of the graphene sample and gap capacitance in series

Z_0 represents the characteristic impedance of the APC-7 connectors (50 Ω)

According to Equation 7.4, although the relaxation time is several femtoseconds, the real part of the surface impedance of macroscale graphene is six orders of magnitude larger than its imaginary part in the microwave frequency range. Therefore, we can deduce Z_g from $Re(Z_{total})$. However, by considering the TEM polarization of the incident field and the ring shape of the graphene, we have

$$Z_g = \frac{\ln(r_{out}/r_{in})R_\square}{2\pi} = \frac{\ln(r_{out}/r_{in})}{2\pi\sigma} \tag{7.6}$$

where r_{out} and r_{in} are the inner radius of the outer conductor and the outer radius of the inner conductor, respectively. Thus, using Equations 7.5 and 7.6, the sheet resistance ($R_\square$) and surface conductivity (σ) of the graphene under test can be extracted.

7.2.1.3 Measurement Results

With the TEM normal incident wave, S-parameter measurement is then performed to characterize the transmission property of different graphene samples. Figure 7.18a,b compares the measured S parameters for our three sets of samples. Figure 7.18c,d presents the corresponding $R_\square$ and σ of each sample. The extracted results of

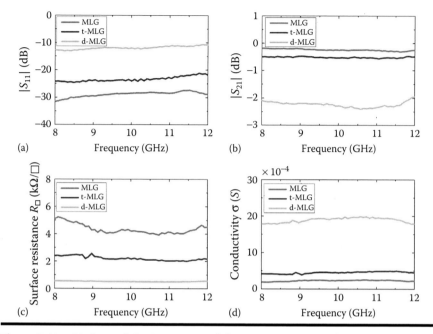

Figure 7.18 (a) S_{11} and (b) S_{21} measured for MLG, t-MLG, and d-MLG on Teflon samples; (c) surface resistance and (d) surface conductivity extracted from measured S parameters.

different graphene samples are also compared in Table 7.2. Our results of graphene surface impedance agree well with those measured by electrode [15] and CPW [23] structure-based methods.

Figure 7.18b shows that the insertion loss of MLG and t-MLG are less than 1 dB, whereas the insertion loss of d-MLG is more than 2 dB, which is led by its much higher surface conductivity. This indicates that doped multilayer graphene may have a potential application in shielding of electromagnetic waves.

From Table 7.2, we can see that the sheet resistance and surface conductivity of t-MLG are very close to half and twice those of MLG, respectively. The reason

Table 7.2 Comparison of Extracted Parameters for Different Graphene Samples

	MLG	*t-MLG*	*d-MLG*
$R_{\square}$ (KΩ/$\square$)	4.35	2.21	0.528
σ ($\times 10^{-4}$ S)	2.31	4.53	19.0
τ (ps) Kubo formula	0.0109	0.0102	0.0282
τ (ps) Einstein relation	0.0101	0.00915	0.0262

is that the sheet resistance of a graphene film should be constant and all layers act nearly independently when the number of stacking layers is less than 4 [29], and the electric properties of each graphene layer in t-MLG are similar to those of MLG. Therefore, t-MLG can be regarded as two MLGs connected in parallel. Moreover, in terms of the doping effect, the conductivity of d-MLG is approximately eight times higher than that of MLG, when the doping density of d-MLG is nearly 10 times of the doping density of MLG. The dependence of the conductivity on doping density n is expected to be $\sigma \propto n^a$ with $a = 1$ for charged impurities and $a < 1$ for short-range and ripple scattering [30]. Scattering other than by charged impurities will dominate at large n [31]. The results obtained from Kubo formula are consistent with the results obtained from Einstein relation ($\tau = \hbar \sigma (\pi/n)^{1/2}/(e^2 v_F)$) and the results demonstrated in [15]. The main reason for the difference between these two methods is that the model used in Einstein relation is simpler than that of the Kubo formula.

7.2.1.4 Conclusion

In summary, a noncontact method to characterize the surface conductivity of graphene at microwave frequency by using APC-7 coaxial connectors is proposed. After that, the scattering rate of graphene on Teflon substrate is calculated by combining the surface conductivity with Fermi energy extracted from Raman spectrum. Three sets of graphene samples (MLG, double-layer graphene, and d-MLG) are used for the testing. Through Raman spectrum analysis, it is also found that the Teflon substrate has a minimized interaction with the graphene layers on top compared with SiO_2 substrate. To avoid large contact resistance in the traditional conductivity measurement, a transmission characterization method is utilized. Thanks to the structure of APC-7 connectors and the calibration process, the de-embedding procedure has been avoided as well, which makes the proposed measurement method not only sufficiently accurate but also simpler than the rectangular waveguide method.

7.2.2 Absorber

With the increasing integration of active components and rapid development of wireless electronics, the electromagnetic radiation pollution is becoming a serious problem. Advanced materials and structures for absorbing microwave energy have attracted much more attention than ever before. Previous studies mainly focus on the thickness, bandwidth, and incidence angle sensitivity of absorbers [32–34]. One problem is that most of those available absorbers are nontransparent and cannot be applied when an observation window is needed. In [35], a wideband transparent absorber was proposed. It utilized an Al wire grid, a polyethylene terephthalate (PET) film, and polydimethylsiloxane layers as the reflective layer, the substrate, and absorbing layer, respectively, where the optical transmittance is affected by the Al wire grid.

In [36], a transparent absorber applying indium tin oxide films was proposed. It is based on the principle of Salisbury absorber and has the limitation that the substrate thickness must be a quarter of working wavelength.

With its frequency-independent and tunable resistance, and high optical transmittance, graphene can be a very good solution for the absorbing film in the transparent and tunable microwave absorber. In [37], a wideband absorber was achieved by stacking unpatterned graphene-bearing quartz substrates, and each substrate has a thickness of a quarter of working wavelength. Furthermore, the design freedom will be improved with a patterned graphene film. In this section, we propose a transparent absorber based on the patterned graphene. The substrate and the reflective layer are glass and fluorine-doped tin oxide (FTO) film, respectively. The graphene film, which is transferred on PET and then patterned as periodical patches, serves as the absorbing layer. The proposed absorber is measured by using a rectangular waveguide.

7.2.2.1 Modeling of the Graphene Absorber

The configuration of the absorber, together with the waveguide for measurement, is shown in Figure 7.19. An equivalent circuit for the absorber placed in the waveguide is proposed in Figure 7.20, from which we can get an insight of its absorbance.

In the equivalent circuit, two TLs are used to represent glass and air region above the absorber, respectively. ε_{ri}, k_{zi}, and Z_{ci} are the relative permittivity, the

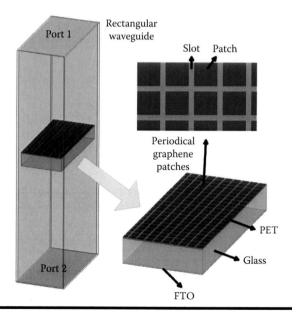

Figure 7.19 **The configuration of the absorber and the waveguide for measurement.**

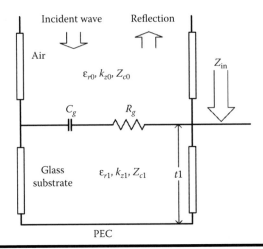

Figure 7.20   The equivalent circuit based on the configuration in Figure 7.19.

longitudinal propagation constant, and the wave impedance of the glass TL and the air TL, respectively ($i = 0$ for the air and $i = 1$ for the glass); t_1 is the thickness of the glass; R_g represents the equivalent resistance of graphene patches; C_g represents the capacitance of slots between graphene patches; and Z_{in} is the input impedance of the absorber. Under the TE_{10} mode of the rectangular waveguide, Z_{ci} is written as

$$Z_{ci} = \frac{\omega \mu_0}{k_{zi}} \qquad (7.7)$$

where:
   ω is the radian frequency
   μ_0 is the permeability of free space

$$k_{zi} = \frac{2\pi}{\lambda_{gi}} = \sqrt{k_0^2 \varepsilon_{ri} - \left(\frac{\pi}{a}\right)^2} \qquad (7.8)$$

where:
   λ_{gi} is the waveguide wavelength of glass substrate ($i = 1$) and air region ($i = 0$)
   a is the long side length of cross section of rectangular waveguide (here WR-62 with $a = 15.8$ mm is used in our measurement)
   $k_0 = \omega/c$ is the wave number in free space, with c being the velocity of light in free space

Because the thickness of PET (0.125 mm) and resistance of FTO (7.5 Ω) are negligible compared with t_1 (2.2 mm) and Z_{c0} (above 200 Ω) respectively, the PET layer is ignored and the FTO layer is set to be a PEC in the equivalent circuit. From the proposed equivalent circuit, the real part and image part of the input impedance Z_{in} of the absorber are obtained as

$$\text{Re}(Z_{in}) = \frac{R_g Y^2}{R_g{}^2 + (Y - X)^2} \tag{7.9}$$

$$\text{Im}(Z_{in}) = \frac{R_g^2 Y - XY(Y - X)}{R_g^2 + (Y - X)^2} \tag{7.10}$$

where:

$$X = \frac{1}{\omega C_g} \tag{7.11}$$

$$Y = Z_{c1} \tan k_{z1} t_1 \tag{7.12}$$

X is the reactance of the capacitance C_g, whereas Y is the reactance of the FTO transformed by the glass TL. When $n\lambda_{g1}/2 < t1 < \lambda_{g1}/4 + n\lambda_{g1}/2$ ($n = 0, 1, 2...$), FTO transformed by the glass TL can be considered as an inductor according to Equation 7.12. It is then connected in parallel with the series of C_g and R_g, so that an resistance-inductance-capacitance (RLC) resonant loop is formed. Because the graphene film itself is resistive at the microwave band, the slots between graphene patches are designed to provide capacitance to compensate the inductance transformed from FTO. By optimizing the dimension of the slots and patches according to the working frequency and the glass substrate thickness, we can get $\text{Re}(Z_{in}) \approx Z_{c0}$ and $\text{Im}(Z_{in}) \approx 0$, so that all the incident energy can go into port 1 due to the impedance matching between the air region and the absorber. At the same time little energy goes through port 2 due to the low resistance of FTO film; most of the incident energy is reflected by the FTO film and absorbed again by the resistive graphene film.

When $\lambda_{g1}/4 + n\lambda_{g1}/2 < t_1 < \lambda_{g1}/2 + n\lambda_{g1}/2$ ($n = 0, 1, 2,...$), the impedance of FTO transformed by the glass substrate is capacitive. The graphene film can be etched as periodical loop rings or periodical crosses so as to provide remarkable series inductance to compensate the capacitance transformed from FTO. Specially, when $t_1 = \lambda_{g1}/4 + n\lambda_{g1}/2$ ($n = 0, 1, 2,...$), the graphene film can serve as the absorbing layer without patterning. At this time, Y is infinite large, $X = 0$, and $Z_{in} = R_g$. That is the available Salisbury graphene absorber. Our proposed graphene absorber is the extension of the available graphene absorber, where the graphene film is patterned to provide enough capacitance/inductance at the microwave band which the unpatterned graphene film cannot provide. Therefore, it can be used with the substrate with an arbitrary thickness.

7.2.2.2 Fabrication and Measurement

A glass substrate whose thickness is 2.2 mm and relative permittivity is 6 is chosen for our experiment. In our example, we choose 12.6 GHz as the desired working frequency. At this frequency, $\lambda_{g1} = 10.2$ mm and $0 < t_1 < \lambda_{g1}/4$, Y is positive, so we pattern the graphene film as periodical patches with slots.

In order to insert the sample into the waveguide for the measurement, small gaps between the edges of the sample and the waveguide walls are left. For our sample, the dimension is measured to be 15.4 mm × 7.3 mm, while the cross section of waveguide is 15.8 mm × 7.9 mm. These gaps result in several parasitic parameters, which affect the accuracy of the equivalent circuit. For example, in the FTO layer and graphene layer, the gaps introduce series capacitance; between the FTO layer and the graphene layer, the gaps introduce fringing capacitance. These parasitic parameters are difficult to obtain using analytical formula.

To better fine-tune the proposed absorber, 3D full-wave simulation is implemented, in which the gaps, the thickness of PET, and the sheet resistance of FTO are taken into account for the accuracy. By varying the sheet resistance, slot width, and side length of the graphene patches, the desired absorption frequency 12.6 GHz is reached when the sheet resistance of graphene film is 46 Ω, slot width 30 μm, and side length 720 μm. The simulated S parameters are shown in Figure 7.22 and the absorption coefficient, which is defined as $A = 1-|S_{11}|^2-|S_{21}|^2$, is shown in Figure 7.23.

MLG film is grown on copper foil using the CVD method. The graphene film is first coated with a hot release tape. Then the copper foil is etched by $FeCl_3$ etchant and the hot release tape/graphene film is left. This film is diluted by DI water, and then transferred onto the PET film. Later, this hot release tape/graphene/PET film is heated at a temperature of 100°C for one minute to remove the hot release tape, and leave graphene on the PET. The Raman spectrum of MLG film on PET is shown in Figure 7.21a. From the two clear characteristic peaks (G band at 1585 cm⁻¹, 2D band at 2695 cm⁻¹), it is demonstrated that the MLG on PET has a good quality. To get a desired sheet resistance, the transfer process is repeated five times to fabricate multilayer graphene film, and this multilayer graphene film is fumed by concentrated nitric acid later, until the sheet resistance decreases to about 46 Ω. The multilayer graphene film on PET is then etched by infrared laser with 1070 nm wavelength. The infrared laser has a resolution of 30 μm. It etches graphene film periodically, leaving the periodical patches with a side length of 720 μm. The optical transmittance of the graphene/PET film is measured using a ultraviolet–visible spectrophotometer and is around 80% in the optical light range, as shown in Figure 7.21b. The fabricated sample and the microgram of the etched periodical graphene patches are shown in Figure 7.21c,d, respectively.

The sample is measured in a rectangular waveguide (WR-62) whose working frequency is from 11.9 to 18 GHz (Ku band). A piece of lossless foam is stuffed into the waveguide to support the sample, as shown in Figure 7.21e. Before measurement,

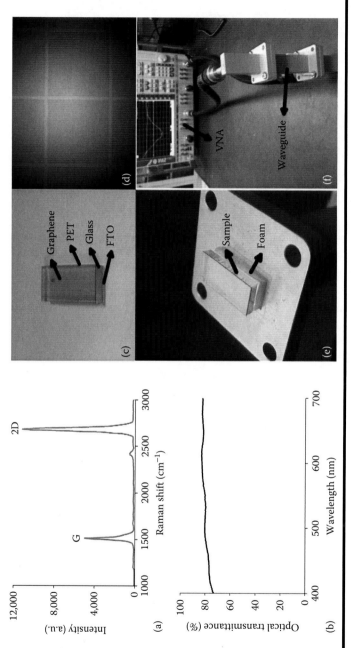

Figure 7.21 (a) The Raman spectrum of MLG on PET, (b) the optical transmittance of the fabricated absorbing film, (c) the fabricated sample, (d) the micrograph of periodical graphene patches, (e) the sample under test, and (f) the full configuration of the measurement.

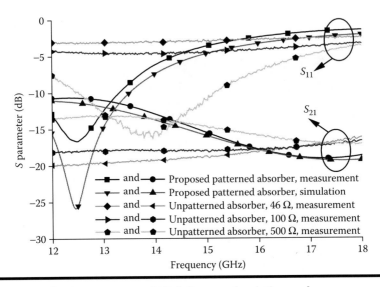

Figure 7.22 The S parameters of 3D full-wave simulation and measurement.

the rectangular waveguide is connected with a VNA and calibrated using the TRL method. The full configuration of the measurement is shown in Figure 7.21f.

The measured scattering parameters are shown in Figure 7.22, in which S_{11} and S_{21} show a very good agreement with the 3D full-wave simulation results. The absorption coefficient is obtained from measured scattering parameters, as shown in Figure 7.23, and the peak absorption occurs at 12.6 GHz and 90% of the incident energy can be absorbed at this frequency.

The proposed absorber is also compared with the available Salisbury graphene absorber. For this purpose, unpatterned graphene/PET films, which have sheet resistances of 46, 100, and 500 Ω, respectively, are fabricated. The absorbers applying these unpatterned films are measured. The scattering parameters and absorption coefficients are shown in Figures 7.22 and 7.23, respectively. We can find that when the sheet resistances of an unpatterned film are 46 and 100 Ω, S_{11} is quite large and the absorption coefficient is smaller. This can be illustrated by the impedance mismatch. Z_{c0} is calculated to be decreasing from 310 to 220 Ω in the frequency range from 12 to 18 GHz by using Equation 7.9. For the unpatterned films, the equivalent resistance R_g is calculated by

$$R_g = R_s \times \frac{b}{a} \tag{7.13}$$

where:
R_s is the sheet resistance of graphene film
a (15.8 mm) and b (7.9 mm) are the long side length and the short side length of the cross section of the waveguide, respectively

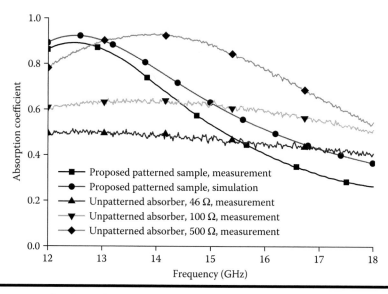

Figure 7.23 **The absorption coefficients of 3D full-wave simulation and measurement.**

When the sheet resistances of the graphene/PET film are 46 and 100 Ω, R_g will be 23 and 50 Ω, respectively. In these cases, R_g is far less than Z_{c0}, which causes the mismatch between R_g and Z_{c0}, and results in low absorption. However, when the sheet resistance of the graphene/PET film increases to 500 Ω, R_g will be 250 Ω, which is very close to Z_{c0}. In this case, we can find that the peak absorption frequency occurs at 14 GHz, and more than 90% energy is absorbed due to better impedance matching.

The comparison between the results of the patterned sample and those of the unpatterned samples demonstrates that by etching the graphene film, it can provide enough capacitance/inductance at the microwave band to compensate the inductance/capacitance transformed from the reflective layer. Therefore, it can be used with the substrate with an arbitrary thickness.

7.2.2.3 Conclusion

In this section, a transparent absorber based on the patterned graphene film is theoretically illustrated, practically fabricated, and experimentally demonstrated. Its measurement result is in good agreement with 3D full-wave simulation result and shows peak absorption as high as 90%. Furthermore, this study shows that by etching the graphene film, a remarkable capacitive/inductive surface can be obtained at the microwave band, which cannot be achieved by the unpatterned graphene film itself. This is helpful to overcome the application difficulties of graphene films at the microwave band.

References

1. D. Wang, X. C. Wei, J. B. Zhang et al., Back lobe reduction of patch antenna by using high-impedance surface, *IET International Radar Conference*, Hangzhou, China, October 14-16, 2015.
2. X. C. Wei, Y. L. Xu, N. Meng et al., A non-contact graphene surface scattering rate characterization method at microwave frequency by combining Raman spectroscopy and coaxial connectors measurement, *Carbon*, 77, 53–58, 2014.
3. C. L. Holloway, E. F. Kuester, J. A. Gordon et al., An overview of the theory and applications of metasurfaces: the two dimensional equivalent of metasurfaces, *IEEE Antennas Propagat. Mag.*, 54(2), 10–35, 2012.
4. D. Sievenpiper, L. Zhang, R. F. J. Broas et al., High-impedance electromagnetic surfaces with a forbidden frequency band, *IEEE Trans. Microwave Theory Tech.*, 47(11), 2059–2074, 1999.
5. F. Yang and Y. T. Sammi, Reflection phase characterizations of the EBG ground plane for low profile wire antenna application, *IEEE Trans. Antennas Propagt.*, 51(10), 2691–2703, 2003.
6. D. J. Kern, D. H. Werner, A. Monorchio et al., The design synthesis of multiband artificial magnetic conductors using high impedance frequency selective surfaces, *IEEE Trans. Antennas Propagt.*, 53(1), 8–17, 2005.
7. T. Kamgaing and O. M. Ramahi, A novel power plane with integrated simultaneous switching noise mitigation capability using high impedance surface, *IEEE Microw. Wireless Compon. Lett.*, 13(1), 21–23, 2003.
8. S. Clavijo, R. E. Diaz, and W. E. McKinzie, Design methodology for Sievenpiper high-impedance surfaces: An artificial magnetic conductor for positive gain electrically small antennas, *IEEE Trans. Antennas Propagat.*, 51(10), 2678–2690, 2003.
9. O. Luukkonen, C. Simovski, G. Granet et al., Simple and accurate analytical model of planar grids and high-impedance surfaces, comprising metal strips or patches, *IEEE Trans. Antennas Propagat.*, 56(6), 1624–1632, 2008.
10. L. Lin, G. Cheng, W. Y. Yin et al., Shielding cover effects on the RF performance of LDMOSFET power amplifier for WCDMA application, *Asia-Pacific Microwave Conference Proceedings*, Seoul, Korea, November 5-8, 2013.
11. M. P. Robinson, T. M. Benson, C. Christopoulos et al., Analytical formulation for the shielding effectiveness of enclosures with apertures, *IEEE Trans. Electromagn. Compat.*, 40(3), 240–248, 1998.
12. R. Azaro, S. Caorsi, M. Donelli et al., Evaluation of the effects of an external incident electromagnetic wave on metallic enclosures with rectangular apertures, *Microwave and Optical Technology Letters*, 28, 289–293, 2001.
13. D. J. S. Gomez, V. M. Garcia, and M. A. Alvarez, A grounded MoM-based spatial Green's function technique for the analysis of multilayered circuits in rectangular shielded enclosures, *IEEE Trans. Microwave Theory Tech.*, 59(3), 533–541, 2011.
14. A. C. Cangellaris and V. I. Okhmatovski, Novel closed-form Green's function in shielded planar layered media, *IEEE Trans. Microwave Theory Tech.*, 48(12), 2225–2232, 2000.
15. J. S. Gomez-Diaz and J. Perruisseau-Carrier, Microwave to THz properties of graphene and potential antenna applications, *International Symposium on Antennas & Propagation*, Nagoya, Japan, October 29–November 2, 2012, pp. 239–242.

16. D. Yi, X. C. Wei, Y. L. Xu et al., Graphene-silicon diode loaded patch antenna, *International Microwave Workshop Series on Advanced Materials and Processes for RF and THz Applications*, Suzhou, China, July 1-3, 2015.

17. Y. L. Xu, X. C. Wei, and E. P. Li, Three-dimensional tunable frequency selective surface based on vertical graphene micro-ribbons, *J. Electromagnet Wave*, 29(16), 2130–2138, 2015.

18. C. P. Yen, C. Argyropoulos, and A. Alu, Terahertz antenna phase shifters using integrally-gated graphene transmission-lines, *IEEE Trans. Antennas Propagat.*, 61(4), 1528–1537, 2013.

19. G. W. Hanson, Dyadic Green's functions and guided surface waves for a surface conductivity model of graphene, *J. Appl. Phys.*, 103(6), 064302, 2008.

20. Y. W. Tan, Y. Zhang, K. Bolotin et al., Measurement of scattering rate and minimum conductivity in graphene, *Phys. Rev. Lett.*, 99(24), 2007.

21. A. K. M. Newaz, Y. S. Puzyrev, B. Wang et al., Probing charge scattering mechanisms in suspended graphene by varying its dielectric environment, *Nat. Commun.*, 3, 734, 2012.

22. M. Dragoman, D. Neculoiu, A. Cismaru et al., Coplanar waveguide on graphene in the range 40 MHz–110 GHz, *Appl. Phys. Lett.*, 99(3), 033112, 2011.

23. H. S. Skulason, H. V. Nguyen, A. Guermoune et al., 110 GHz measurement of large-area graphene integrated in low-loss microwave structures, *Appl. Phys. Lett.*, 99(15), 153504, 2011.

24. Y. Khatami, H. Li, C. Xu, and K. Banerjee, Metal-to-multilayer-graphene contact-part i: contact resistance modeling, *IEEE Trans. Electron Devices*, 59(9), 2444–2452, 2012.

25. J. S. Gomez-Diaz, J. Perruisseau-Carrier, P. Sharma et al., Non-contact characterization of graphene surface impedance at micro and millimeter waves, *J. Appl. Phys.*, 111(11), 114908, 2012.

26. L. Hao, J. Gallop, S. Goniszewski, O. Shaforost, N. Klein, R. Yakimova, Non-contact method for measurement of the microwave conductivity of graphene, *Appl. Phys. Lett.*, 103(12), 123103, 2013.

27. S. Das, P. Sudhagar, E. Ito et al., Effect of HNO_3 functionalization on large scale graphene for enhanced tri-iodide reduction in dye-sensitized solar cells, *J. Mater. Chem.*, 22(38), 20490–20497, 2012.

28. V. P. Gusynin, S. G. Sharapov, and J. P. Carbotte, Magneto-optical conductivity in graphene, *J. Phys: Condens Matter.*, 19(2), 026222, 2007.

29. X. Li, Y. Zhu, W. Cai et al., Transfer of large-area graphene films for high-performance transparent conductive electrodes, *Nano Lett.*, 9(12), 4359–4363, 2009.

30. J. H. Chen, C. Jang, A. Adam et al., Charged-impurity scattering in graphene, *Nat. Phys.*, 4(5), 377–381, 2008.

31. E. H. Hwang, A. Adam, and S. S. Das, Carrier transport in two-dimensional graphene layers, *Phys. Rev. Lett.*, 98(18), 2007.

32. R. L. Fante and M. T. McCormack, Reflection properties of the Salisbury screen, *IEEE Trans. Antennas Propagat.*, 36(10), 1443–1454, 1988.

33. J. Yuan and Z. Shen, A thin and broadband absorber using double-square loops, *IEEE Antennas Wireless Propagat. Lett.*, 6(11), 388–391, 2007.

34. F. Costa, A. Monorchio, and G. Manara, Analysis and design of ultra thin electromagnetic absorbers comprising resistively loaded high impedance surfaces, *IEEE Trans. Antennas Propagat.*, 58(5), 1551–1558, 2010.

35. T. Jang, H. Youn, Y. J. Shin et al., Transparent and flexible polarization-independent microwave broadband absorber, *ACS Photonics,* 1(3), 279–284, 2014.
36. K. Takizawa and O. Hashimoto, Transparent wave absorber using resistive thin film at V-band frequency, *IEEE Trans. Microwave Theory Tech.,* 47(7), 1137–1141, 1999.
37. B. Wu, H. M. Tuncer, M. Naeem et al., Experimental demonstration of a transparent graphene millimetre wave absorber with 28% fractional bandwidth at 140 GHz, *Sci. Rep.* [Online], vol. 4, available: http://www.nature.com/srep/2014/140219/srep04130/abs/srep04130.html#supplementary-information.

Index